BASICS OF CLASSICAL MECHANICS

BASICS OF CLASSICAL MECHANICS

Alimuddin Khan

ANMOL PUBLICATIONS PVT. LTD.
NEW DELHI - 110 002 (INDIA)

ANMOL PUBLICATIONS PVT. LTD.
H.O.: 4374/4B, Ansari Road, Daryaganj,
New Delhi-110 002 (India)
Ph.: 23278000, 23261597
B.O.: No. 1015, Ist Main Road, BSK IIIrd Stage
IIIrd Phase, IIIrd Block,
Bangalore - 560 085 (India)
Visit us at: www.anmolpublications.com

Basics of Classical Mechanics

First Edition, 2008
ISBN 978-81-261-3728-2

PRINTED IN INDIA

Printed at Mehra Offset Press, Delhi.

Contents

Preface

Basics of Classical Mechanics cover the field of classical mechanics at the undergraduate/postgraduate level. This book is intended for students who have studied some mechanics in an introductory physics course. With unusual clarity, this text coves most of the topics normally found in books at this level." It is self-contained, and combines thoroughness with lucidity of presentation. Classical mechanics is developed with particular emphasis on exploring the classical limit of what is sometimes referred to as quantum statistical mechanics.

It will be useful not only to graduate students of physics, but to students of related disciplines like chemistry, biology, and chemical engineering. With such a wide audience in mind, it includes in the opening chapter a presentation of the basic notions of Mechanics. The book contains a number of appendices that will be of interest to the relatively advanced students as well as to specialists.

Author

Chapter 1

Introduction

Classical mechanics is the study of the *motion* of bodies in accordance with the general principles first enunciated by Sir Isaac Newton in his *Philosophiae Naturalis Principia Mathematica*,commonly known as the *Principia.* Classical mechanics was the first branch of Physics to be discovered, and is the foundation upon which all other branches of Physics are built. Moreover, classical mechanics has many important applications in other areas of science, such as Astronomy, Chemistry, Geology, and Engineering . Classical mechanics is also of great significance outside the realm of science.

After all, the sequence of events leading to the discovery of classical mechanics—starting with the ground-breaking work of Copernicus, continuing with the researches of Galileo, Kepler, and Descartes, and culminating in the monumental achievements of Newton—involved the complete overthrow of the Aristotelian picture of the Universe, which had previously prevailed for more than a millennium, and its replacement by a recognizably modern picture in which humankind no longer played a privileged role.

In our investigation of classical mechanics we shall study many different types of motion, including:

Translational motion: Motion by which a body shifts from one point in space to another .

Rotational motion: Motion by which an extended body changes orientation, with respect to other bodies in space, without changing position .

Oscillatory motion: Motion which continually repeats in time with a fixed period.

Circular motion: Motion by which a body executes a circular orbit about another fixed body.

Of course, these different types of motion can be combined: for instance, the motion of a properly bowled bowling ball consists of a combination of translational and rotational motion, whereas wave propagation is a combination of translational and oscillatory motion. Furthermore, the types of motion are not entirely distinct: *e.g.*, circular motion contains elements of both rotational and oscillatory motion. We shall also study *statics*: *i.e.*, the subdivision of mechanics which is concerned with the forces that act on bodies *at rest* and in equilibrium. Statics is obviously of great importance in civil engineering: for instance, the principles of statics were used to design the building in which this lecture is taking place, so as to ensure that it does not collapse.

mks Units

The first principle of any exact science is *measurement*. In mechanics there are three fundamental quantities which are subject to measurement:

1. Intervals in space: *i.e.*, lengths.
2. Quantities of inertia, or mass, possessed by various bodies.
3. Intervals in time.

Any other type of measurement in mechanics can be reduced to some combination of measurements of these three quantities.

Each of the three fundamental quantities—*length, mass*, and *time*—is measured with respect to some convenient standard. The system of units currently used by all scientists, and most engineers, is called the *mks system*—after the first initials of the names of the units of length, mass, and time, respectively, in this system: *i.e.*, the *meter*, the *kilogram*, and the *second*.

The mks unit of length is the *meter* (symbol m), which was formerly the distance between two scratches on a platinum-iridium alloy bar kept at the International Bureau of Metric Standard in Sèvres, France, but is now defined as the distance occupied by 1, 650, 763.73 wavelengths of light of the orange-red spectral line of the isotope Krypton 86 in vacuum.

The mks unit of mass is the *kilogram* (symbol kg), which is defined as the mass of a platinum-iridium alloy cylinder kept at the International Bureau of Metric Standard in Sèvres, France. The mks unit of time is the *second* (symbol s), which was formerly defined in terms of the Earth's rotation, but is now defined as the time for 9, 192, 631, 770 oscillations associated with the transition between the two hyperfine levels of the ground state of the isotope Cesium 133.

In addition to the three fundamental quantities, classical mechanics also deals with *derived quantities*, such as velocity, acceleration, momentum, angular momentum, *etc*. Each of these derived quantities can be reduced to some particular combination of length, mass, and time. The mks units of these derived quantities are, therefore, the corresponding combinations of the mks units of length, mass, and time. For instance, a velocity can be reduced to a length divided by a time. Hence, the mks units of velocity are meters per second:

$$[v]=\frac{[L]}{[T]}=ms^{-1}$$

Here, *v* stands for a velocity, *L* for a length, and *T* for a time, whereas the operator [...] represents the units, or *dimensions*, of the quantity contained within the brackets. Momentum can be reduced to a mass times a velocity. Hence, the mks units of momentum are kilogram-meters per second:

$$[p]=[M][v]=\frac{[M][L]}{[T]}=\text{kgms}^{-1}.$$

Here, p stands for a momentum, and M for a mass. In this manner, the mks units of all derived quantities appearing in classical dynamics can easily be obtained.

STANDARD PREFIXES

mks units are specifically designed to conveniently describe those motions which occur in everyday life. Unfortunately, mks units tend to become rather unwieldy when dealing with motions on very small scales or very large scales . In order to help cope with this problem, a set of standard prefixes has been devised, which allow the mks units of length, mass, and time to be modified so as to deal more easily with very small and very large quantities: these prefixes are specified in Table, Thus, a *kilometer* (km) represents 10^3 m, a *nanometer* (nm) represents 10^{-9} m, and a *femtosecond* (fs) represents 10^{-15}s. The standard prefixes can also be used to modify the units of derived quantities.

Table: Standard Prefixes

Factor	Prefix	Symbol	Factor	Prefix	Symbol
10^{18}	exa-	E	10^{-1}	deci-	D
10^{15}	Peta-	P	10^{-2}	centi-	C
10^{12}	Tera-	T	10^{-3}	micro-	M
10^{9}	Giga-	G	10^{-6}	micro-	μ
10^{6}	mega-	M	10^{-9}	nano-	N
10^{3}	Kilo-	K	10^{-12}	pico-	P
10^{2}	hecto-	H	10^{-15}	femto-	F
10^{1}	deka-	Da	10^{-18}	atto-	A

OTHER UNITS

The mks system is not the only system of units in existence. Unfortunately, the obsolete cgs system and the even more obsolete fps system are still in use today, although their continued employment is now *strongly discouraged* in science and engineering. Conversion between different systems of units is, in principle, perfectly straightforward, but, in practice, a frequent source of error. Witness, the recent loss of the Mars

Climate Orbiter because the engineers who designed its rocket engine used fps units whereas the NASA mission controllers employed mks units. Table specifies the various conversion factors between mks, cgs, and fps units. Note that, rather confusingly, a pound is a unit of force, rather than mass. Additional non-standard units of length include the inch (1ft = 12in), the yard (1ya = 3ft), and the mile (1mi = 5, 280ft). Additional non-standard units of mass include the ton (in the US, 1ton = 2,0001b; in the UK, 1to = 2,2401b), and the metric ton (1tonne = 1,000kg). Finally, additional non-standard units of time include the minute (1min = 60s), the hour (1 hr = 3,600s), the day (1da = 86, 400s), and the year (1yr = 365.26 da = 31, 558, 464s).

Table: Conversion factors

$1\text{cm} = 10^{-2}\text{m}$
$1\text{g} = 10^{-3}\text{kg}$
1ft = 0.3048m
1lb = 4.448N
1slug = 14.59kg

PRECISION AND SIGNIFICANT FIGURES

We are expected to perform calculations to a relative accuracy of 1 per cent: *i.e.*, to *three significant figures*. Since rounding errors tend to accumulate during lengthy calculations, the easiest way in which to achieve this accuracy is to perform all intermediate calculations to *four significant figures*, and then to round the final result down to three significant figures. If one of the quantities in our calculation turns out to the the small difference between two much larger numbers, then we may need to keep *more* than four significant figures.

Incidentally, we are strongly urged to use *scientific notation* in *all* of our calculations: the use of non-scientific notation is generally a major source of error in this course. If our calculators are capable of operating in a mode in which *all* numbers are displayed in scientific form then we are advised to perform our calculations in this mode.

DIMENSIONAL ANALYSIS

Length, mass, and time are three *fundamentally different* quantities which are measured in three completely independent units. It, therefore, makes no sense for a prospective law of physics to express an equality between a length and a mass. In other words, the example law

$$m = l,$$

where m is a mass and l is a length, cannot possibly be correct. One easy way of seeing that Eq. is invalid, is to note that this equation is dependent on the adopted system of units: *i.e.*, if $m = l$ in mks units, then $m \neq l$ in fps units, because the conversion factors which must be applied to the left- and right-hand sides differ. Physicists hold very strongly to the assumption that the laws of physics possess *objective reality*: in other words, the laws of physics are the same for all observers.

One immediate consequence of this assumption is that a law of physics must take the same form in all possible systems of units that a prospective observer might choose to employ. The only way in which this can be the case is if all laws of physics are *dimensionally consistent*: *i.e.*, the quantities on the left- and right-hand sides of the equality sign in any given law of physics must have the same dimensions. A dimensionally consistent equation naturally takes the same form in all possible systems of units, since the same conversion factors are applied to both sides of the equation when transforming from one system to another.

As an example, let us consider what is probably the most famous equation in physics:

$$E = mc^2$$

Here, E is the energy of a body, is its mass, and is the velocity of light in vacuum. The dimensions of energy are [M] $[L^2]/[T^2]$, and the dimensions of velocity are $[L]/[T]$. Hence, the dimensions of the left-hand side are [M] $[L^2]/[T^2]$, whereas the dimensions of the right-hand side are [M] $([L]/[T])^2$ = [M] $[L^2]/[T^2]$. It follows that equation is indeed dimensionally consistent. Thus, $E = mc^2$ holds good in mks units, in cgs units, in fps units, and in any other sensible set of units. Had Einstein proposed $E = mc$, or $E = mc^3$, then his error would have been immediately apparent to other physicists, since these prospective laws are not dimensionally consistent. In fact, $E = mc^2$ represents the *only* simple, dimensionally consistent way of combining an energy, a mass, and the velocity of light in a law of physics.

The last comment leads naturally to the subject of *dimensional analysis*: *i.e.*, the use of the idea of dimensional consistency to *guess* the forms of simple laws of physics. It should be noted that dimensional analysis is of fairly limited applicability, and is a poor substitute for analysis employing the actual laws of physics; nevertheless, it is occasionally useful. Suppose that a special effects studio wants to film a scene in which the Leaning Tower of Pisa topples to the ground. In order to achieve this, the studio might make a scale model of the tower, which is 1m tall, and then film the model falling over. The only problem is that the resulting footage would look completely unrealistic, because the model tower would fall over too quickly. The studio could easily fix this problem by slowing the film down.

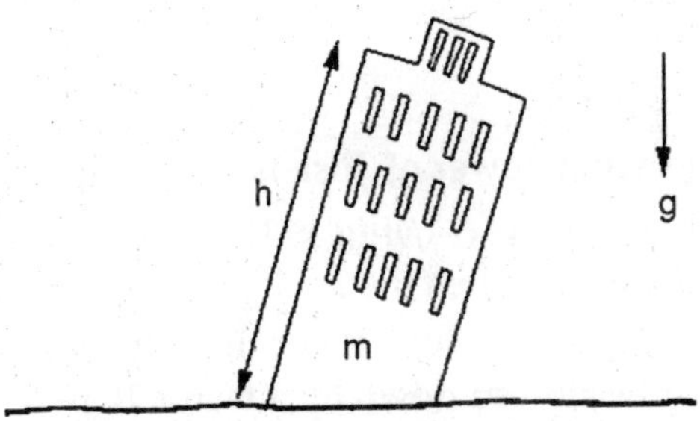

Fig. The Leaning Tower of Pisa

Although, at this stage, we do not know how to apply the laws of physics to the problem of a tower falling over, we can, at least, make some educated guesses as to what factors the time t_f required for this process to occur depends on. In fact, it seems reasonable to suppose that depends principally on the mass of the tower, m, the height of the tower, h, and the acceleration due to gravity, g. In other words,

$$t_f = C\, m^x h^y g^z$$

where C is a dimensionless constant, and x, y, and z are unknown exponents. The exponents x, y, and z can be determined by the requirement that equation be dimensionally consistent. Incidentally, the dimensions of an acceleration are $[L]/[T^2]$. Hence, equating the dimensions of both sides of equation, we obtain

$$[T] = [M]^x [L]^y \left(\frac{[L]}{[T^2]} \right)^z$$

We can now compare the exponents of $[L]$, $[M]$, and $[T]$ on either side of the expression: these exponents must all match in order for equation to be dimensionally consistent. Thus,

$$\begin{aligned} 0 &= y + z, \\ 0 &= x, \\ 1 &= -2z \end{aligned}$$

It immediately follows that $x = 0$, $y = 1/2$, and $z = -1/2$. Hence,

$$t_f = C\sqrt{\frac{h}{g}}$$

Now, the actual tower of Pisa is approximately 100m tall. It follows that since $t_f \,\alpha\, \sqrt{h}$ (g is the same for both the real and the model tower) then the 1m high model tower falls over a factor of $\sqrt{100/1} = 10$ times faster than the real tower. Thus, the film must be slowed down by a factor 10 in order to make it look realistic.

Problem: Farmer Jones has recently brought a 40 acre field and wishes to replace the fence surrounding it. Given that the field is square, what length of fencing (in meters) should Farmer Jones purchase? Incidentally, 1 acre equals 43,560 square feet.

Solution : If 1 acre equals 43,560ft^2 and 1 ft equals 0.3048m then

$$1 \text{ acre} = 43560 \times (0.3048)^2 = 4.047 \times 10^3 m^2$$

Thus, the area of the field in mks units is

$$A = 40 \times 4.047 \times 10^3 = 1.619 \times 10^5 m^2$$

Now, a square field with sides of length l has an area $A = l^2$ and a circumference $D = 4l$. Hence, $D = \sqrt[4]{A}$. It follows that the length of the fence is

$$D = 4 \times \sqrt{1.619 \times 10^5} = 1.609 \times 10^3 m$$

Problem: The recommended tire pressure in a Honda Civic is 28 psi (pounds per square inch). What is this pressure in atmospheres (1 atmosphere is 10^5 Nm^{-2})?

Solution: First, 28 pounds per square inch is the same as 28 $\times (12)^2 = 4032$ pounds per square foot (the standard fps unit of pressure). Now, 1 pound equals 4.448 Newtons (the standard SI unit of force), and 1 foot equals 0.3048m. Hence,

$$P = 4032 \times (4.448)/(0.3048)^2 = 1.93 \times 10^5 Nm^{-2}$$

It follows that 28 psi is equivalent to 1.93 atmospheres.

Problem: The speed of sound v in a gas might plausibly depend on the pressure p, the density ρ, and the volume V of the gas. Use dimensional analysis to determine the exponents x, y, and z in the formula

$$v = Cp^x p^y V^z$$

where C is a dimensionless constant. Incidentally, the mks units of pressure are kilograms per meter per second squared.

Solution : Equating the dimensions of both sides of the equation, we obtain

$$\frac{[L]}{[T]} = \left(\frac{[M]}{[T^2[L]]}\right)^x \left(\frac{[M]}{[L^3]}\right)^y [L^3]^z$$

A comparison of the exponents of $[L]$, $[M]$, and $[T]$ on either side of the expression yields

$$
\begin{aligned}
1 &= -x - 3y + 3z \\
0 &= x + y, \\
-1 &= -2x
\end{aligned}
$$

The third equation immediately gives $x = 1/2$; the second equation then yields $y = -1/2$; finally, the first equation gives $z = 0$. Hence,

$$v = C\sqrt{\frac{p}{\rho}}.$$

Chapter 2

Newton's Laws of Motion

In his *Principia,* Newton reduced the basic principles of mechanics to three laws:

Every body continues in its state of rest, or uniform motion in a straight line, unless compelled to change that state by forces impressed upon it. The change of motion of an object is proportional to the force impressed upon it, and is made in the direction of the straight line in which the force is impressed. To every action there is always opposed an equal reaction; or, the mutual actions of two bodies upon each other are always equal and directed to contrary parts.

These laws are known as Newton's first law of motion, Newton's second law of motion, and Newton's third law of motion, respectively.

NEWTON'S FIRST LAW OF MOTION

Newton's first law was actually discovered by Galileo and perfected by Descartes (who added the crucial proviso "in a straight line"). This law states that if the motion of a given body is not disturbed by external influences then that body moves with *constant* velocity. In other words, the displacement r of the body as a function of time t can be written

$$r = r_0 + vt$$

where r_0 and v are *constant* vectors. The body's trajectory is a *straight-line* which passes through point r_0 at time $t = 0$

and runs parallel to v. In the special case in which $v = 0$ the body simply remains at rest.

Nowadays, Newton's first law strikes us as almost a statement of the obvious. However, in Galileo's time this was far from being the case. From the time of the ancient Greeks, philosophers—observing that objects set into motion on the Earth's surface eventually come to rest—had concluded that the natural state of motion of objects was that they should remain at rest. Hence, they reasoned, any object which moves does so under the influence of an external influence, or *force,* exerted on it by some other object.

It took the genius of Galileo to realize that an object set into motion on the Earth's surface eventually comes to rest under the influence of frictional forces, and that if these forces could somehow be abstracted from the motion then it would continue forever.

NEWTON'S SECOND LAW OF MOTION

Newton used the word "motion" to mean what we nowadays call *momentum*. The momentum P of a body is simply defined as the product of its mass m and its velocity v :

i.e., $p = m\,v$

Newton's second law of motion is summed up in the equation

$$\frac{dp}{dt} = f$$

where the vector f represents the net influence, or force, exerted on the object, whose motion is under investigation, by other objects. For the case of a object with *constant* mass, the above law reduces to its more conventional form $f = m.a$

In other words, the net force exerted on a given object by other objects equals the product of that object's mass and its acceleration. Of course, this law is entirely devoid of content unless we have some independent means of quantifying the forces exerted between different objects.

HOOKE'S LAW

One method of quantifying the force exerted on an object is via *Hooke's law*. This law—discovered by the English scientist Robert Hooke in 1660—states that the force f exerted by a coiled spring is directly proportional to its *extension* Δx. The extension of the spring is the difference between its actual length and its natural length (*i.e.*, its length when it is exerting no force).

The force acts parallel to the axis of the spring. Obviously, Hooke's law only holds if the extension of the spring is sufficiently small. If the extension becomes too large then the spring deforms permanently, or even breaks. Such behaviour lies beyond the scope of Hooke's law.

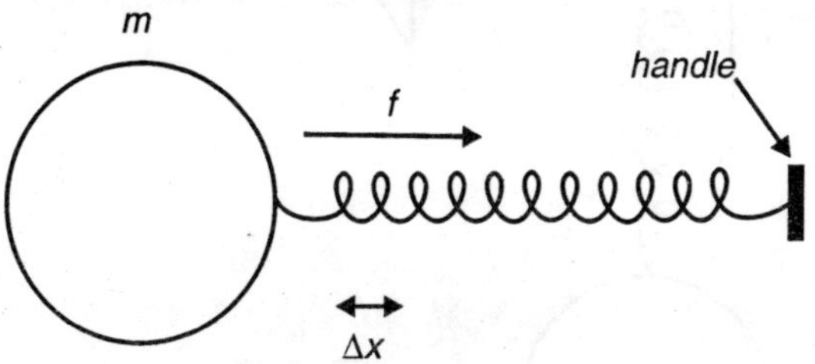

Fig. *Hooke's law*

figure illustrates how we might use Hooke's law to quantify the force we exert on a body of mass when we pull on the handle of a spring attached to it. The magnitude of the force is proportional to the extension of the spring: twice the extension means twice the force. As shown, the direction of the force is towards the spring, parallel to its axis (assuming that the extension is positive).

The magnitude of the force can be quantified in terms of the critical extension required to impart a unit acceleration (1m/sec^2)to a body of unit mass (1 kg). According to equation, the force corresponding to this extension is 1 *newton*. Here, a newton (symbol N) is equivalent to a kilogram-meter per second-squared, and is the mks unit of force. Thus, if the critical extension corresponds to a force of 1 N then half the critical extension corresponds to a force of 0.5N, and so on. In this manner, we can quantify both the direction and magnitude of the force we exert, by means of a spring, on a given body.

Suppose that we apply two forces, f_1 and f_2, acting in different directions, to a body of mass m by means of two springs. The body accelerates as if it were subject to a single force which is the *vector sum* of the individual forces f_1 and f_2. It follows that the force f appearing in Newton's second law of motion, equation, is the *resultant* of all the external forces to which the body whose motion is under investigation is subject.

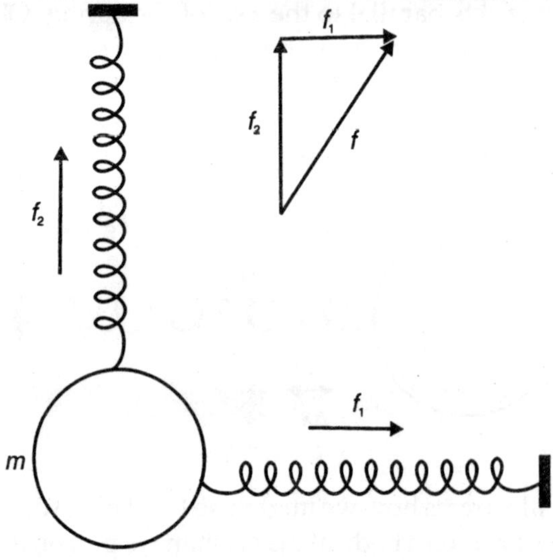

Fig. Addition of forces

Suppose that the resultant of all the forces acting on a given body is *zero*. In other words, suppose that the forces acting on the body exactly balance one another. According to Newton's second law of motion, equation, the body does not accelerate: *i.e.*, it either remains at rest or moves with uniform velocity in a straight line. It follows that Newton's first law of motion applies not only to bodies which have no forces acting upon them but also to bodies acted upon by exactly balanced forces.

NEWTON'S THIRD LAW OF MOTION

Suppose, for the sake of argument, that there are only two bodies in the Universe. Let us label these bodies *a* and *b*.

Suppose that body *b* exerts a force f_{ab} on body *a*. According to Newton's third law of motion, body must exert an *equal and opposite* force $f_{ba} = -f_{ab}$ on body . Thus, if we label f_{ab} the "action" then, in Newton's language, f_{ba} is the equal and opposed "reaction".

Suppose, now, that there are many objects in the Universe (as is, indeed, the case). According to Newton's third law, if object j exerts a force f_{ij} then object i must exert an equal and opposite force $f_{ji} = -f_{ij}$ on object j. It follows that all of the forces acting in the Universe can ultimately be grouped into equal and opposite action-reaction pairs. Note, incidentally, that an action and its associated reaction always act on *different* bodies.

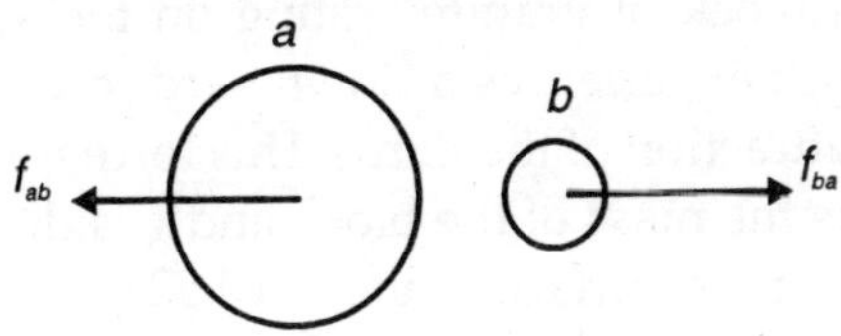

Fig. Newton's third law

Why do we need Newton's third law? Actually, it is almost a matter of common sense. Suppose that bodies *a* and *b* constitute an *isolated* system. If $f_{ba} \neq -f_{ab}$ then this system exerts a *non-zero net force* $f = f_{ab} + f_{ba}$ on itself, without the aid of any external agency. It will, therefore, accelerate forever under its own steam. We know, from experience, that this sort of behaviour does not occur in real life.

MASS AND WEIGHT

The terms *mass* and *weight* are often confused with one another. However, in physics their meanings are quite distinct.

A body's mass is a measure of its *inertia*: *i.e.*, its reluctance to deviate from uniform straight-line motion under the influence of external forces. According to Newton's second law, equation, if two objects of differing masses are acted upon by forces of the same magnitude then the resulting acceleration of the larger mass is less than that of the smaller mass. In other words, it is more difficult to force the larger mass to deviate from its preferred state of uniform motion in a straight line. Incidentally, the mass of a body is an intrinsic property of that body, and, therefore, does not change if the body is moved to a different place.

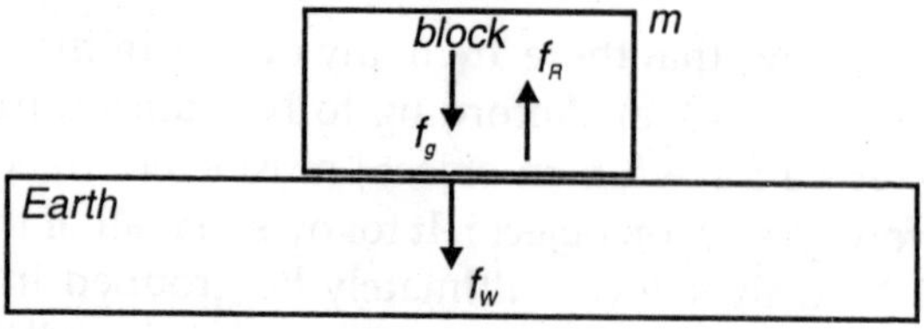

Fig. Weight

Imagine a block of granite resting on the surface of the Earth. The block experiences a downward force f_g due to the gravitational attraction of the Earth. This force is of magnitude *mg*, where *m* is the mass of the block and *g* is the acceleration due to gravity at the surface of the Earth. The block transmits this force to the ground below it, which is supporting it, and, thereby, preventing it from accelerating downwards.

In other words, the block exerts a downward force f_w, of magnitude m_g, on the ground immediately beneath it. We usually refer to this force (or the magnitude of this force) as the *weight* of the block. According to Newton's third law, the ground below the block exerts an upward reaction force f_R on the block. This force is also of magnitude *mg*. Thus, the net force acting on the block is $f_g + f_R = 0$, which accounts for the fact that the block remains stationary.

Where, we might ask, is the equal and opposite reaction to the force of gravitational attraction f_g exerted by the Earth on the block of granite? It turns out that this reaction is exerted at the centre of the Earth. In other words, the Earth attracts the block of granite, and the block of granite attracts the Earth by an equal amount. However, since the Earth is far more massive than the block, the force exerted by the granite block at the centre of the Earth has no observable consequence.

So far, we have established that the weight *W* of a body is the magnitude of the downward force it exerts on any object which supports it. Thus, $W = mg$, where is the mass of the body and g is the local acceleration due to gravity. Since weight is a force, it is measured in newtons. A body's weight is location dependent, and is not, therefore, an intrinsic property of that body. For instance, a body weighing 10N on the surface of the Earth will only weigh about 3.8N on the surface of Mars, due to the weaker surface gravity of Mars relative to the Earth.

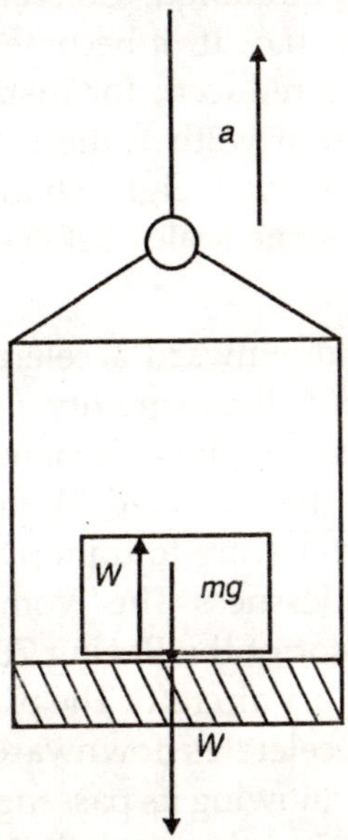

Fig. Weight in an elevator

Consider a block of mass m resting on the floor of an elevator. Suppose that the elevator is accelerating upwards with acceleration a. How does this acceleration affect the weight of the block? Of course, the block experiences a downward force mg due to gravity. Let W be the weight of the block: by definition, this is the size of the downward force exerted by the block on the floor of the elevator.

From Newton's third law, the floor of the elevator exerts an upward reaction force of magnitude W on the block. Let us apply Newton's second law, equation, to the motion of the block. The mass of the block is m, and its upward acceleration is. Furthermore, the block is subject to two forces: a downward force mg due to gravity, and an upward reaction force W. Hence,

$$W - mg = ma$$

This equation can be rearranged to give

$$W = m\,(g+a).$$

Clearly, the upward acceleration of the elevator has the effect of increasing the weight W of the block: for instance, if the elevator accelerates upwards at $g = 9.81\ m/sec^2$ then the weight of the block is doubled. Conversely, if the elevator accelerates downward (*i.e.*, if a becomes negative) then the weight of the block is reduced: for instance, if the elevator accelerates downward at $g/2$then the weight of the block is halved. Incidentally, these weight changes could easily be measured by placing some scales between the block and the floor of the elevator.

Suppose that the downward acceleration of the elevator matches the acceleration due to gravity: *i.e.*, $a = -g$. In this case, $W = 0$. In other words, the block becomes weightless! This is the principle behind the so-called "Vomit Comet" used by NASA's Johnson Space Centre to train prospective astronauts in the effects of weightlessness. The "Vomit Comet" is actually a KC-135 (a predecessor of the Boeing 707 which is typically used for refueling military aircraft). The plane typically ascends to 30,000ft and then accelerates downwards at g(*i.e.*, drops like a stone) for about 20s, allowing its passengers to feel the effects of weightlessness during this period. All of the weightless scenes in the film *Apollo 11* were shot in this manner.

Suppose, finally, that the downward acceleration of the elevator exceeds the acceleration due to gravity: *i.e.*$a < -g$. In this case, the block acquires a negative weight! What actually happens is that the block flies off the floor of the elevator and slams into the ceiling: when things have settled down, the block exerts an upward force (negative weight) $|W|$on the ceiling of the elevator.

STRINGS, PULLEYS, AND INCLINES

Consider a block of mass m which is suspended from a fixed beam by means of a string. The string is assumed to be light (*i.e.*, its mass is negligible compared to that of the block) and inextensible (*i.e.*, its length increases by a negligible amount because of the weight of the block). The string is clearly being stretched, since it is being pulled at both ends by the block and the beam. Furthermore, the string must be being pulled by oppositely directed forces of the same magnitude, otherwise it would accelerate greatly (given that it has negligible inertia). By Newton's third law, the string exerts oppositely directed forces of equal magnitude, T, on both the block and the beam.

These forces act so as to oppose the stretching of the string: *i.e.*, the beam experiences a downward force of magnitude T, whereas the block experiences an upward force of magnitude T. Here, T is termed the *tension* of the string. Since T is a force, it is measured in newtons. Note that, unlike a coiled spring, a string can never possess a negative tension, since this would imply that the string is trying to push its supports apart, rather than pull them together.

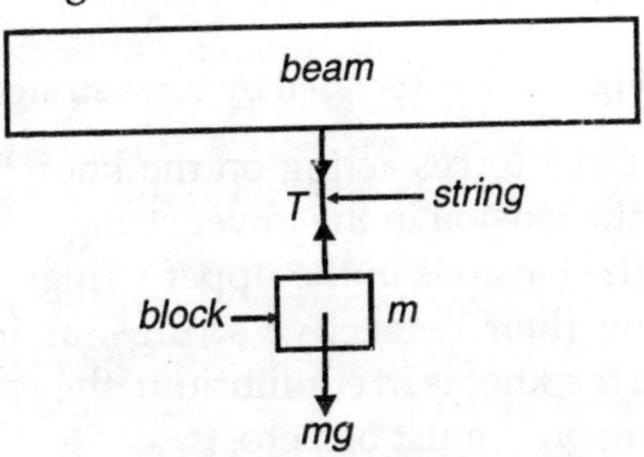

Fig. Block suspended by a string

Let us apply Newton's second law to the block. The mass of the block is m, and its acceleration is zero, since the block is assumed to be in equilibrium. The block is subject to two forces, a downward force mg due to gravity, and an upward force T due to the tension of the string. It follows that

$$T - mg = 0$$

In other words, in equilibrium, the tension T of the string equals the weight mg of the block.

A slightly more complicated example in which a block of mass m is suspended by three strings. The question is what are the tensions, T, T_1, and T_2, in these strings, assuming that the block is in equilibrium? Using analogous arguments to the previous case, we can easily demonstrate that the tension T in the lowermost string is mg. The tensions in the two uppermost strings are obtained by applying Newton's second law of motion to the knot where all three strings meet.

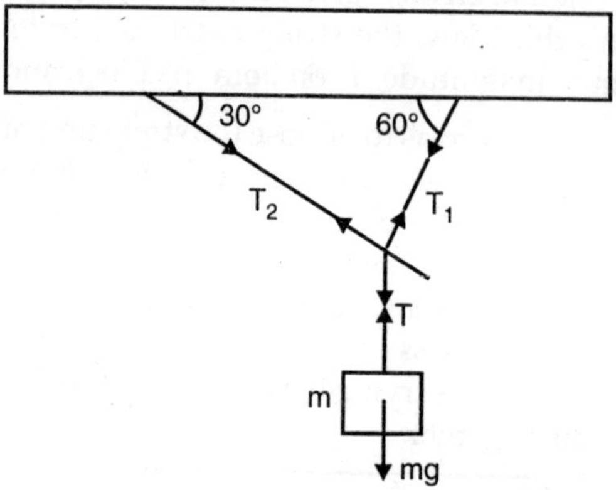

Fig. Block suspended by three strings

There are three forces acting on the knot: the downward force T due to the tension in the lower string, and the forces T_1 and T_2 due to the tensions in the upper strings. The latter two forces act along their respective strings, as indicate in the diagram. Since the knot is in equilibrium, the vector sum of all the forces acting on it must be zero.

Consider the horizontal components of the forces acting on the knot. Let components acting to the right be positive, and *vice versa*. The horizontal component of tension T is zero, since this tension acts straight down. The horizontal component of tension T_1 is $T_1 \cos 60° = T_1/2$, since this force subtends an angle of with respect to the horizontal. Likewise, the horizontal component of tension T_2 is $-T_2 \cos 30° = -\sqrt{3}T_2/2$. Since the knot does not accelerate in the horizontal direction, we can equate the sum of these components to zero:

$$\frac{T_1}{2} - \frac{\sqrt{3}T_2}{2} = 0$$

Consider the vertical components of the forces acting on the knot. Let components acting upward be positive, and *vice versa*. The vertical component of tension T is $-T = -mg$, since this tension acts straight down. The vertical component of tension T_1 is $T_1 \sin 60° = \sqrt{3}T_1/2$, since this force subtends an angle of 60° with respect to the horizontal. Likewise, the vertical component of tension T_2 is $T_2 \sin 30° = T_2/2$. Since the knot does not accelerate in the vertical direction, we can equate the sum of these components to zero:

$$-mg + \frac{\sqrt{3}T_1}{2} + \frac{T_2}{2} = 0$$

Finally, Equation $T_1 = \frac{\sqrt{3}mg}{2}, T_2 = \frac{mg}{2}$

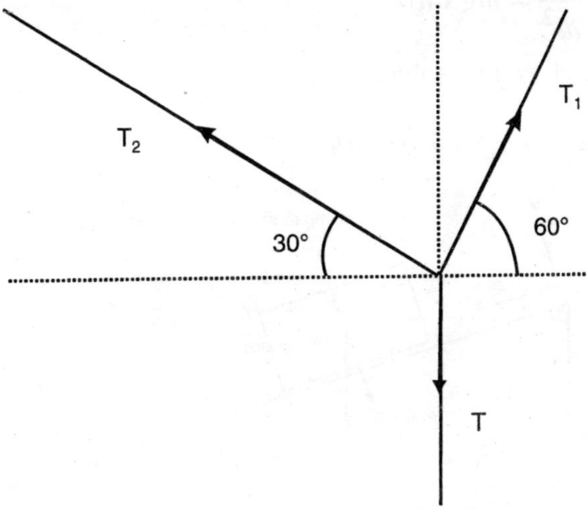

Detail of Figure:

Consider a block of mass m sliding down a smooth frictionless incline which subtends an angle θ to the horizontal. The weight mg of the block is directed vertically downwards. However, this force can be resolved into components $mg \cos\theta$, acting perpendicular (or normal) to the incline, and $mg \sin\theta$, acting parallel to the incline. Note that the reaction of the incline to the weight of the block acts *normal* to the incline, and only matches the *normal component* of the weight: *i.e.*, it is of magnitude $mg\cos\theta$.

This is a general result: the reaction of any unyielding surface is always locally normal to that surface, directed outwards (away from the surface), and matches the normal component of any inward force applied to the surface. The block is clearly in equilibrium in the direction normal to the incline, since the normal component of the block's weight is balanced by the reaction of the incline. However, the block is subject to the unbalanced force $mg\sin\theta$ in the direction parallel to the incline, and, therefore, accelerates down the slope. Applying Newton's second law to this problem, we obtain

$$m\frac{d^2x}{dt^2} = mg\sin\theta$$

which can be solved to give

$$x = x_0 + v_0t + \frac{1}{2}g\sin\theta t^2$$

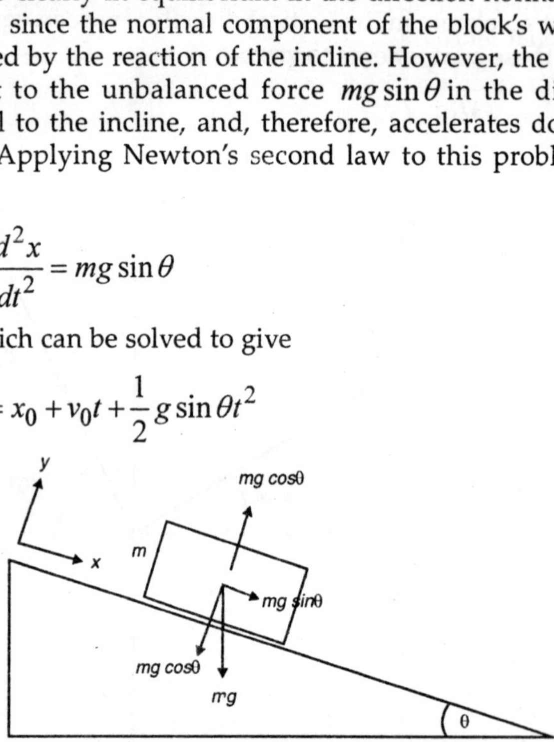

Fig. Block sliding down an incline

In other words, the block accelerates down the slope with acceleration $g\sin\theta$. Note that this acceleration is *less* than the full acceleration due to gravity, *g*. In fact, if the incline is fairly gentle (*i.e.*, if θ is small) then the acceleration of the block can be made *much less* than *g*. This was the technique used by Galileo in his pioneering studies of motion under gravity—by diluting the acceleration due to gravity, using inclined planes, he was able to obtain motion sufficiently slow for him to make accurate measurements using the crude time-keeping devices available in the 17th Century.

Consider two masses, m_1 and m_2, connected by a light inextensible string. Suppose that the first mass slides over a smooth, frictionless, horizontal table, whilst the second is suspended over the edge of the table by means of a light frictionless pulley. Since the pulley is light, we can neglect its rotational inertia in our analysis.

Moreover, no force is required to turn a frictionless pulley, so we can assume that the tension T of the string is the same on either side of the pulley. Let us apply Newton's second law of motion to each mass in turn. The first mass is subject to a downward force $m_1 g$, due to gravity. However, this force is completely canceled out by the upward reaction force due to the table. The mass m_1 is also subject to a horizontal force T, due to the tension in the string, which causes it to move *rightwards* with acceleration

$$a = \frac{T}{m_1}$$

The second mass is subject to a downward force $m_2 g$, due to gravity, plus an upward force T due to the tension in the string. These forces cause the mass to move *downwards* with acceleration

$$a = g - \frac{T}{m_2}$$

Now, the rightward acceleration of the first mass must match the downward acceleration of the second, since the string which connects them is inextensible. Thus, equating the previous two expressions, we obtain

$$T = \frac{m_1 + m_2}{m_1 + m_2} g$$

$$a = \frac{m_2}{m_1 + m_2} g$$

Note that the acceleration of the two coupled masses is *less* than the full acceleration due to gravity, g, since the first mass contributes to the inertia of the system, but does not contribute to the downward gravitational force which sets the system in motion.

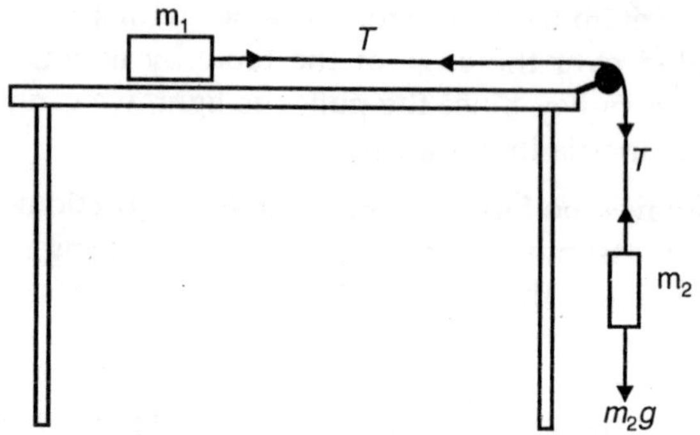

Fig. Block sliding over a smooth table, pulled by a second block

Consider two masses, m_1 and m_2, connected by a light inextensible string which is suspended from a light frictionless pulley. Let us again apply Newton's second law to each mass in turn. Without being given the values of m_1 and m_2, we cannot determine beforehand which mass is going to move upwards. Let us *assume* that mass m_1 is going to move upwards: if we are wrong in this assumption then we will simply obtain a negative acceleration for this mass. The first mass is subject to an upward force T, due to the tension in the string, and a downward force m_1g, due to gravity. These forces cause the mass to move *upwards* with acceleration

$$a = \frac{T}{m_1} - g$$

The second mass is subject to a downward force m_2g, due to gravity, and an upward force T, due to the tension in the string. These forces cause the mass to move *downward* with acceleration

$$a = g - \frac{T}{m_2}$$

Now, the upward acceleration of the first mass must match the downward acceleration of the second, since they are connected by an inextensible string. Hence, equating the previous two expressions, we obtain

$$T = \frac{2m_1 m_2}{m_1 + m_2} g$$

$$a = \frac{m_2 - m_1}{m_1 + m_2} g$$

As expected, the first mass accelerates upward (*i.e.* $a > 0$) if $m_2 > m_1$, and *vice versa.* Note that the acceleration of the system is *less* than the full acceleration due to gravity, g, since both masses contribute to the inertia of the system, but their weights partially cancel one another out. In particular, if the two masses are almost equal then the acceleration of the system becomes very much less than g.

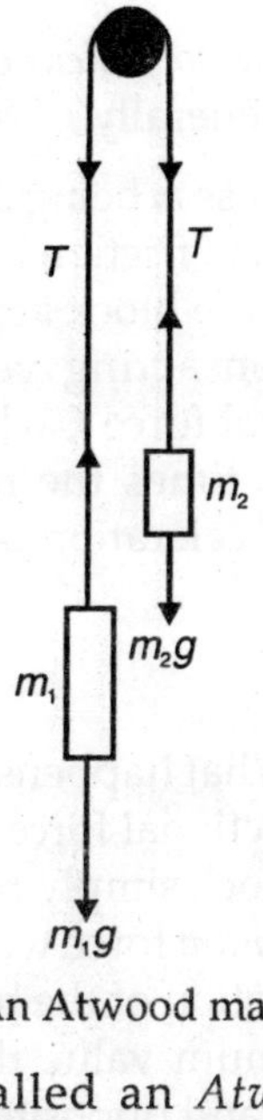

Fig. An Atwood machine

Incidentally it is called an *Atwood machine,* after the eighteenth Century English scientist George Atwood, who used it to "slow down" free-fall sufficiently to make accurate observations of this phenomena using the primitive time-keeping devices available in his day.

FRICTION

When a body slides over a rough surface a frictional force generally develops which acts to impede the motion. Friction, when viewed at the microscopic level, is actually a very complicated phenomenon. Nevertheless, physicists and engineers have managed to develop a relatively simple empirical law of force which allows the effects of friction to be incorporated into their calculations. This law of force was first proposed by Leonardo da Vinci, and later extended by Charles Augustin de Coulomb. The frictional force exerted on a body sliding over a rough surface is proportional to the *normal reaction* R_n at that surface, the constant of proportionality depending on the nature of the surface. In other words,

$$f = \mu R_n$$

where μ is termed the *coefficient of (dynamical) friction*. For ordinary surfaces, μ is generally of order unity.

Consider a block of mass *m* being dragged over a horizontal surface, whose coefficient of friction is μ, by a horizontal force *F*. The weight $W = mg$ of the block acts vertically downwards, giving rise to a reaction acting vertically upwards. The magnitude of the frictional force *f*, which impedes the motion of the block, is simply μ times the normal reaction $R = mg$. Hence, $f = \mu mg$. The acceleration of the block is, therefore,

$$a = \frac{F - f}{m} = \frac{F}{m} - \mu g$$

assuming that $F > f$. What happens if $F < f$: *i.e.*, if the applied force *F* is less than the frictional force *f*? In this case, common sense suggests that the block simply remains at rest. Hence, $f = \mu mg$ is actually the *maximum* force which friction can generate in order to impede the motion of the block. If the applied force *F* is less than this maximum value then the applied force is canceled out by an equal and opposite frictional force, and the block remains stationary. Only if the applied force exceeds the maximum frictional force does the block start to move.

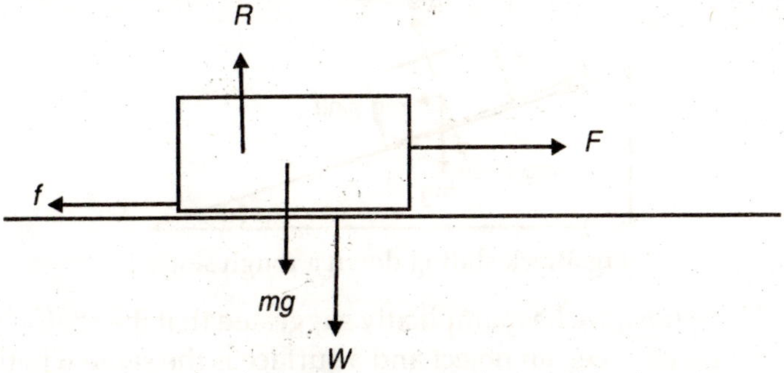

Fig. Friction

Consider a block of mass *m* sliding down a rough incline (coefficient of friction μ) which subtends an angle θ to the horizontal. The weight *mg* of the block can be resolved into components $mg\cos\theta$, acting normal to the incline, and $mg\sin\theta$, acting parallel to the incline. The reaction of the incline to the weight of the block acts *normally* outwards from the incline, and is of magnitude $mg\cos\theta$.

Parallel to the incline, the block is subject to the downward gravitational force, and the upward frictional force *f*(which acts to prevent the block sliding down the incline). In order for the block to move, the magnitude of the former force must exceed the maximum value of the latter, which is μ time the magnitude of the normal reaction, or $\mu mg\cos\theta$. Hence, the condition for the weight of the block to overcome friction, and, thus, to cause the block to slide down the incline, is

$$mg\sin\theta > \mu mg\cos\theta,$$

or

$$\tan\theta > \mu$$

In other words, if the slope of the incline exceeds a certain critical value, which depends on μ, then the block will start to slide. Incidentally, the formula suggests a fairly simple way of determining the coefficient of friction for a given object sliding over a particular surface. Simply tilt the surface gradually until the object just starts to move: the coefficient of friction is simply the tangent of the critical tilt angle (measured with respect to the horizontal).

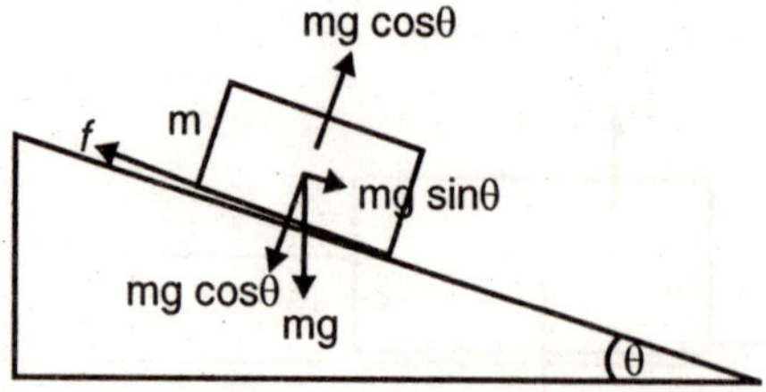

Fig. Block sliding down a rough slope

Up to now, we have implicitly suggested that the coefficient of friction between an object and a surface is the same whether the object remains stationary or slides over the surface. In fact, this is generally not the case. Usually, the coefficient of friction when the object is stationary is slightly *larger* than the coefficient when the object is sliding. We call the former coefficient the *coefficient of static friction,* μ_s, whereas the latter coefficient is usually termed the *coefficient of kinetic (or dynamical) friction,* μ_k.

The fact that $\mu_s > \mu_k$ simply implies that objects have a tendency to "stick" to rough surfaces when placed upon them. The force required to unstick a given object, and, thereby, set it in motion, is μ_s times the normal reaction at the surface. Once the object has been set in motion, the frictional force acting to impede this motion falls somewhat to μ_k times the normal reaction.

FRAMES OF REFERENCE

The laws of physics are assumed to possess *objective reality.* In other words, it is assumed that two independent observers, studying the same physical phenomenon, would eventually formulate *identical* laws of physics in order to account for their observations. Now, two completely independent observers are likely to choose different systems of units with which to quantify physical measurements.

The dimensional consistency of valid laws of physics renders them *invariant* under transformation from one system of units to another. Independent observers are also likely to choose different coordinate systems. For instance, the origins of their separate coordinate systems might differ, as well as the orientation of the various coordinate axes. Are the laws of physics also *invariant* under transformation between coordinate systems possessing different origins, or a different orientation of the various coordinate axes?

Consider the vector equation

$$r = r_1 + r_2$$

Suppose that we shift the origin of our coordinate system, or rotate the coordinate axes. Clearly, in general, the *components* of vectors r, r_1, and r_2 are going to be modified by this change in our coordinate scheme. However, Figure still remains valid. Hence, we conclude that the vector equation also remains valid. In other words, although the individual components of vectors r, r_1, and r_2 are modified by the change in coordinate scheme, the interrelation between these components expressed in Equation remains invariant. This observation suggests that the independence of the laws of physics from the arbitrary choice of the location of the underlying coordinate system's origin, or the equally arbitrary choice of the orientation of the various coordinate axes, can be made manifest by simply writing these laws as interrelations between vectors. In particular, Newton's second law of motion,

$$f = ma$$

is clearly invariant under shifts in the origin of our coordinate system, or changes in the orientation of the various coordinate axes. Note that the quantity m(*i.e.*, the mass of the body whose motion is under investigation), appearing in the equation, is invariant under any changes in the coordinate system, since measurements of mass are completely independent of measurements of distance. We refer to such a quantity as a *scalar* (this is an improved definition). We conclude that valid laws of physics must consist of combinations of scalars and vectors, otherwise they would retain an unphysical dependence on the details of the chosen coordinate system.

Up to now, we have implicitly assumed that all of our observers are *stationary* (*i.e.*, they are all standing still on the surface of the Earth). Let us, now, relax this assumption. Consider two observers, O and O', whose coordinate systems coincide momentarily at $t = 0$. Suppose that observer O is stationary (on the surface of the Earth), whereas observer O' moves (with respect to observer O) with *uniform* velocity v_0. If r represents the displacement of some body P in the stationary observer's frame of reference, at time t, then the corresponding displacement in the moving observer's frame of reference is simply

$$r' = r - v_0 t$$

The velocity of body P in the stationary observer's frame of reference is defined as $v = \frac{dr}{dt}$,

Hence, the corresponding velocity in the moving observer's frame of reference takes the form

$$v' = \frac{dr'}{dt} = v - v_0 .$$

Finally, the acceleration of body P in stationary observer's frame of reference is defined as

$$a = \frac{dv}{dt}$$

whereas the corresponding acceleration in the moving observer's frame of reference takes the form

$$a' = \frac{dv'}{dt} = a$$

Hence, the acceleration of body P is *identical* in both frames of reference.

It is clear that if observer O concludes that body P is moving with constant velocity, and, therefore, subject to zero net force, then observer will agree with this conclusion. Furthermore, if observer O concludes that body P is accelerating, and, therefore, subject to a force a/m, then observer O' will remain in agreement. It follows that Newton's laws of motion are equally valid in the frames of reference of the moving and the stationary observer. Such frames are termed *inertial frames of reference.*

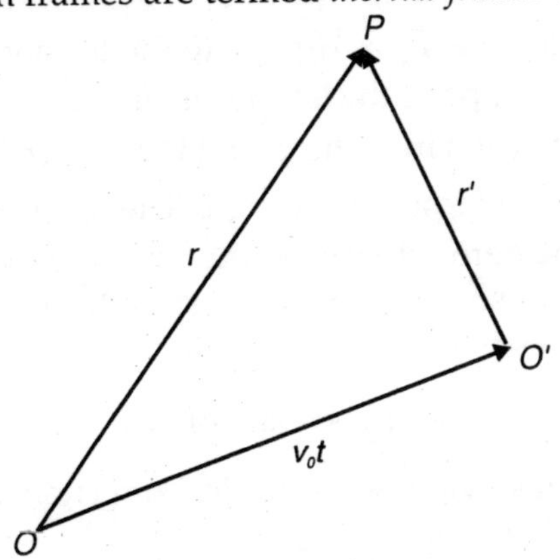

Fig. A moving observer

There are infinitely many inertial frames of reference—within which Newton's laws of motion are equally valid—all moving with *constant* velocity with respect to one another. Consequently, there is no universal standard of rest in physics. Observer might claim to be at rest compared to observer, and *vice versa*: however, both points of view are equally valid. Moreover, there is absolutely no physical experiment which observer O could perform in order to demonstrate that he/she is at rest whilst observer O' is moving. This, in essence, is the principle of *special relativity,* first formulated by Albert Einstein in 1905.

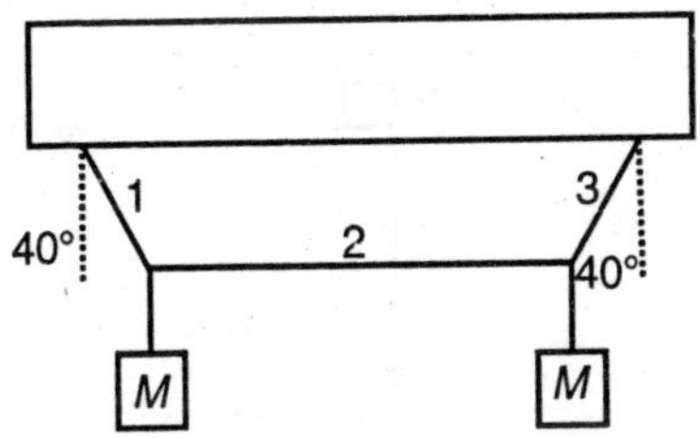

Problem : Consider the diagram. If the system is in equilibrium, and the tension in string 2 is 50N, determine the mass M.

Solution: It follows from symmetry that the tensions in strings 1 and 3 are equal. Let T_1 be the tension in string 1, and T_2 the tension in string 2. Consider the equilibrium of the knot above the leftmost mass. This knot is subject to three forces: the downward force $T_4 = Mg$ due to the tension in the string which directly supports the leftmost mass, the rightward force T_2 due to the tension in string 2, and the upward and leftward force T_1 due to the tension in string 1. The resultant of all these forces must be zero, otherwise the system would not be in equilibrium. Resolving in the horizontal direction (with rightward forces positive), we obtain

$$T_2 - T_1 \sin 40° = 0$$

Likewise, resolving in the vertical direction (with upward forces positive) yields

$$T_1 \cos 40° - T_4 = 0$$

Combining the above two expressions, making use of the fact that $T_4 = Mg$, gives

$$M = \frac{T_2}{g \tan 40°}$$

Finally, since $T_2 = 50N$ and $g = 9.81 m/s^2$, we obtain

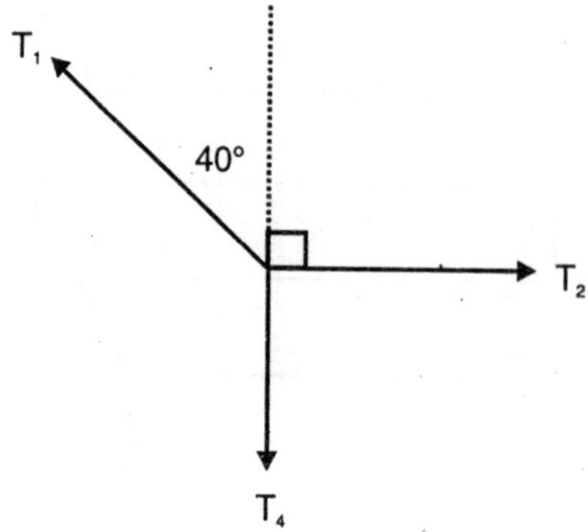

$$M = \frac{50}{9.81 \times 0.8391} = 6.074 kg$$

Problem: Consider the diagram. Suppose that the block, mass $m = 5kg$, is subject to a horizontal force $F = 27N$. What is the acceleration of the block up the (frictionless) slope?

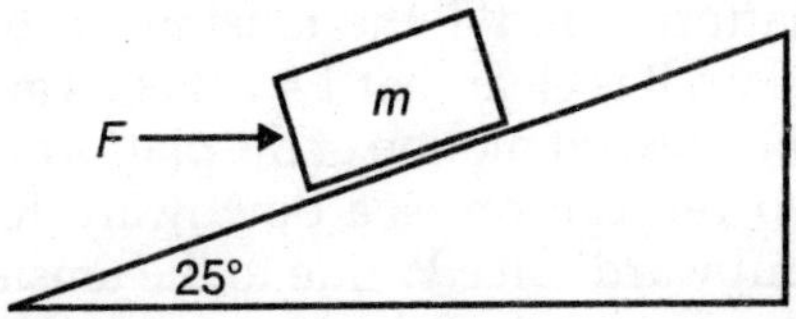

Solution : Only that component of the applied force which is parallel to the incline has any influence on the block's motion: the normal component of the applied force is canceled out by the normal reaction of the incline. The component of the applied force acting up the incline is $F \cos 25°$. Likewise, the component of the block's weight acting down the incline is $mg \sin 25°$. Hence, using Newton's second law to determine the acceleration of the block up the incline, we obtain

$$a = \frac{F\cos 25° - mg\sin 25°}{m}$$

Since $m = 5kg$ and $F = 27N$, we have

$$a = \frac{27 \times 0.9063 - 5 \times 9.81 \times 0.4226}{5} = 0.7483 m/s^2$$

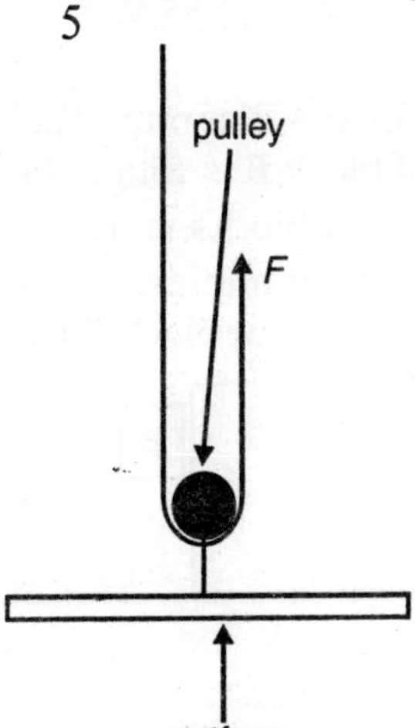

Problem: Consider the diagram. The platform and the attached frictionless pulley weigh a total of 34N. With what force F must the (light) rope be pulled in order to lift the platform at 3.2m/s^2?

Solution : Let W be the weight of the platform, $m = W/g$ the mass of the platform, and T the tension in the rope. From Newton's third law, it is clear that $T = F$. Let us apply Newton's second law to the upward motion of the platform. The platform is subject to two vertical forces: a downward force due to its weight, and an upward force W due to the tension in the rope (the force is $2T$, rather than T, because both the leftmost and rightmost sections of the rope, emerging from the pulley, are in tension and exerting an upward force on the pulley). Thus, the upward acceleration of the platform is

$$a = \frac{2T - W}{m}$$

Since $T = F$ and $m = W / g$, we obtain

$$F = \frac{W(a/g+1)}{2}$$

Finally, given that $W = 34N$ and $a = 3.2m/s^2$, we have

$$F = \frac{34(3.2/9.81+1)}{2} = 22.55N$$

Problem : Consider the diagram. The mass of block A is 75kg and the mass of block B is 15kg. The coefficient of static friction between the two blocks is $\mu = 0.45$. The horizontal surface is frictionless. What minimum force F must be exerted on block A in order to prevent block B from falling?

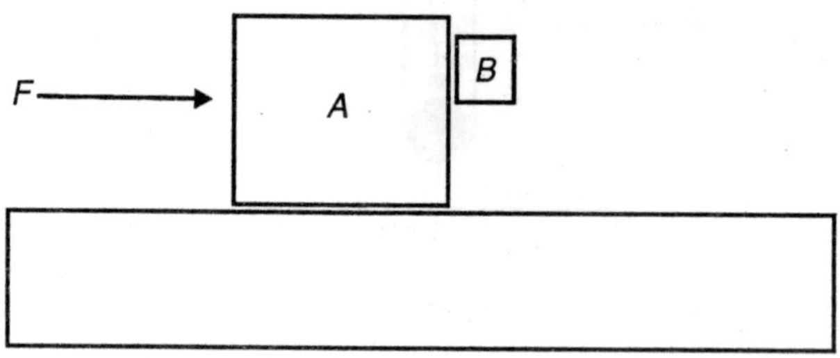

Soluiton: Suppose that block A exerts a rightward force R on block B. By Newton's third law, block B exerts an equal and opposite force on block *A*. Applying Newton's second law of motion to the rightward acceleration *a* of block *B*, we obtain

$$a = \frac{R}{m},$$

where m_B is the mass of block *B*. The normal reaction at the interface between the two blocks is . Hence, the maximum frictional force that block *A* can exert on block *B* is μR. In order to prevent block *B* from falling, this maximum frictional force (which acts upwards) must exceed the downward acting weight, $m_B g$, of the block. Hence, we require

$$\mu R > m_B g$$

or

$$a > \frac{g}{\mu}.$$

Applying Newton's second law to the rightward acceleration *a* of both blocks (remembering that the equal and opposite forces exerted between the blocks cancel one another out), we obtain

$$a = \frac{F}{m_A + m_B},$$

where m_A is the mass of block *A*. It follows that

$$F > \frac{(m_A + m_B)g}{\mu}$$

Since, $m_A = 75kg, m_B = 15kg$ and $\mu = 0.45$, we have

$$F > \frac{(75+15)\times 9.81}{0.45} = 1.962\times 10^3 N$$

Chapter 3

Rotational Motion

ROTATIONAL MOTION OF A RIGID BODY ABOUT A FIXED AXIS

A rigid body is defined as a system of particles, the distance between any pair of which remains unaltered. This is an ideal condition, but objects like a metal block or a log of wood under moderate forces and torques approach very close to being 'rigid' bodies.

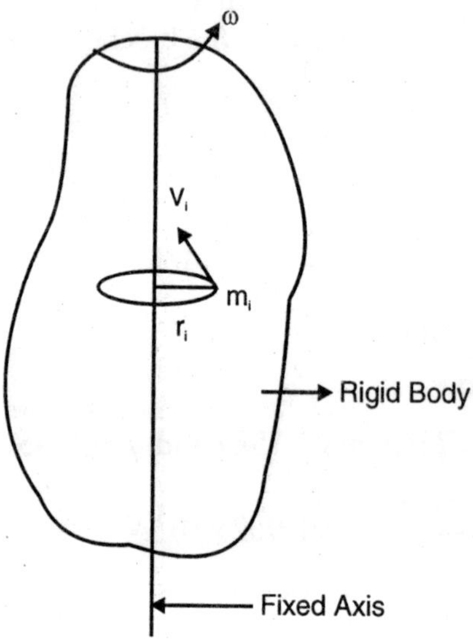

The rotational motion of a rigid body about a fixed axis can be described by introducing concepts of Moment of Inertia, torque, angular momentum etc.

Consider a rigid body rotating about a fixed axis. Let m_i, be the mass of the ith particle, let r_i be its⊥[r] distance from the axis and let vi be its tangential velocity ⊥[r] to radius r_i of the circle in which the particle rotates with same angular velocity as that of the body.

$$\text{Then } (K.E)_i = \frac{1}{2}m_i v_i^2$$

$$= \frac{1}{2}m_i (r_1 w)^2$$

$$= \frac{1}{2}m_i r_1 w^2$$

$$\therefore (K.E)_{body} = \sum_{i=1}^{N}\frac{1}{2}m_i r_i w^2$$

$$= \frac{1}{2}\left(\sum_{i=1}^{N} m_i r_i^2\right) w^2$$

$$\therefore (K.E)_{body} = \frac{1}{2}Iw^2$$

$$\text{where I} = \sum^{i} m_i r_i^2$$

The Quantity I is called the moment of Inertia (M.I) of the rigid body about the given axis. It is defined as the sum over the body of mass of each particle multiplied by the square of its ⊥[r] distance from the axis of rotation. Its unit is 1 Kg × m².

TORQUE ON THE RIGID BODY AXIS

Torque $\overrightarrow{T_1}$ on i[th] particle is defined as

$$\overrightarrow{T_i} = \overrightarrow{r_i} \times \overrightarrow{F_i}$$

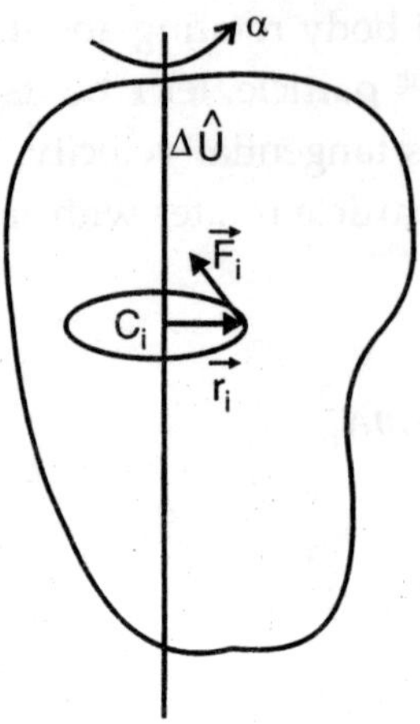

where $\vec{r_i}$: position vector of ith particle from the center c_i of the circle shown.

$\vec{F_i}$: Force on i[th] particle which can be along a tangential direction only.

$$\text{Now } \vec{r_i} \times \vec{F_i} = r_i F_i \widehat{u_i}$$

$$\text{but } F_i = m_i a_i$$

$$= m_i \frac{dv_i}{dt}$$

$$= m_i \frac{d}{dt}(r_1 \omega)$$

$$= m_i r_i \frac{d_\omega}{dt} (\because r_i = \text{constant})$$

$$= m_i r_i \alpha$$

$$\therefore \vec{r_i} \times \vec{F_i} = r_i (m_i r_i \alpha) \hat{u}$$

$$= m_i r_1^2 \alpha \hat{u}$$

$$= m_i r_1^2 \vec{\alpha}$$

$$\therefore \vec{\tau_i} = m_i r_i^2 \vec{\alpha}$$

$$\therefore \vec{\tau}_{body} = \sum_{i=1}^{N} \vec{\tau_i} = \sum_{i=1}^{N} m_i r_i^2 \vec{\alpha}$$

$$\therefore (\vec{\tau})_{body} = \left(\sum_{i=1}^{N} m_i r_i^2 \right) \vec{\alpha}$$

$$\therefore (\vec{\tau})_{body} = 1\vec{\alpha}$$

ANGULAR MOMENTUM OF THE RIGID BODY AXIS

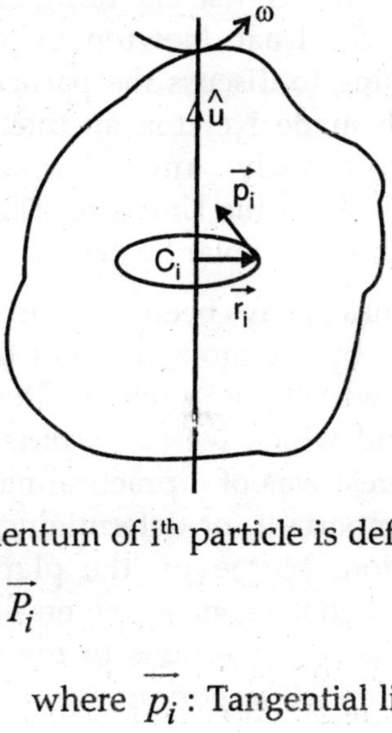

Angular Momentum of ith particle is defined as

$$\vec{L_i} = \vec{r_i} \times \vec{P_i}$$

where $\vec{p_i}$: Tangential linear momentum.

Now, $\vec{r_i} \times \vec{p_i} = r_i p_i \hat{u}$

but $p_i = m_i v_i = m_i r_i \omega$

$$\therefore \vec{r_i} \times \vec{p_i} = r_i (m_i r_i \omega) \hat{u}$$

Chapter 4

Orbital Motion

We have spent this course exploring the theory of motion first outlined by Sir Isaac Newton in his *Principia.* It is, therefore, interesting to discuss the particular application of this theory which made Newton an international celebrity, and which profoundly and permanently changed humankind's outlook on the Universe. This application is, of course, the motion of the Solar System.

Humankind has always been fascinated by the night sky, and, in particular, by the movements of the Sun, the Moon, and the objects which the ancient Greeks called *plantai* ("wanderers"), and which we call *planets.* In ancient times, much of this interest was of a practical nature. The Sun and the Moon were important for determining the calendar, and also for navigation. Moreover, the planets were vital to astrology: *i.e.,* the belief—almost universally prevalent in the ancient world—that the positions of the planets in the sky could be used to foretell important events.

Actually, there were only seven "wandering" heavenly bodies visible to ancient peoples: the Sun, the Moon, and the five planets—Mercury, Venus, Mars, Jupiter, and Saturn. The ancients believed that the stars were fixed to a "celestial sphere" which formed the outer boundary of the Universe. However, it was recognized that the wandering bodies were located *within* this sphere: *e.g.,* because the Moon clearly passes in front of, and blocks the light from, stars in its path. It was also recognized that some bodies were closer to the

Earth than others. For instance, ancient astronomers noted that the Moon occasionally passes in front of the Sun and each of the planets. Moreover, Mercury and Venus can sometimes be seen to transit in front of the Sun.

The first scientific model of the Solar System was outlined by the Greek philosopher Eudoxas of Cnidus. According to this model, the Sun, the Moon, and the planets all execute uniform circular orbits around the Earth—which is fixed, and non-rotating. The order of the orbits is as follows: Moon, Mercury, Venus, Sun, Mars, Jupiter, Saturn—with the Moon closest to the Earth. For obvious reasons, Eudoxas' model became known as the *geocentric* model of the Solar System. Note that orbits are circular in this model for philosophical reasons. The ancients believed the heavens to be the realm of perfection. Since a circle is the most "perfect" imaginable shape, it follows that heavenly objects must execute circular orbits.

A second Greek philosopher, Aristarchus of Samos, proposed an alternative model in which the Earth and the planets execute uniform circular orbits around the Sun—which is fixed. Moreover, the Moon orbits around the Earth, and the Earth rotates daily about a North-South axis. The order of the planetary orbits is as follows: Mercury, Venus, Earth, Mars, Jupiter, Saturn—with Mercury closest to the Sun. This model became known as the *heliocentric* model of the Solar System.

The heliocentric model was generally rejected by the ancient philosophers for three main reasons:

1. If the Earth is rotating about its axis, and orbiting around the Sun, then the Earth must be in motion. However, we cannot "feel" this motion. Nor does this motion give rise to any obvious observational consequences. Hence, the Earth must be stationary.
2. If the Earth is executing a circular orbit around the Sun then the positions of the stars should be slightly different when the Earth is on opposite sides of the Sun. This effect is known as *parallax*. Since no stellar

parallax is observable (at least, with the naked eye), the Earth must be stationary. In order to appreciate the force of this argument, it is important to realise that ancient astronomers did not suppose the stars to be significantly further away from the Earth than the planets. The celestial sphere was assumed to lie just beyond the orbit of Saturn.

3. The geocentric model is far more philosophically attractive than the heliocentric model, since in the former model the Earth occupies a privileged position in the Universe.

The geocentric model was first converted into a proper scientific theory, capable of accurate predictions, by the Alexandrian philosopher Claudius Ptolemy. Basically, Ptolemy acquired and extended the extensive set of planetary observations of his predecessor Hipparchus, and then constructed a geocentric model capable of accounting for them. However, in order to fit the observations, Ptolemy was forced to make some significant modifications to the original model of Eudoxas. Let us discuss these modifications.

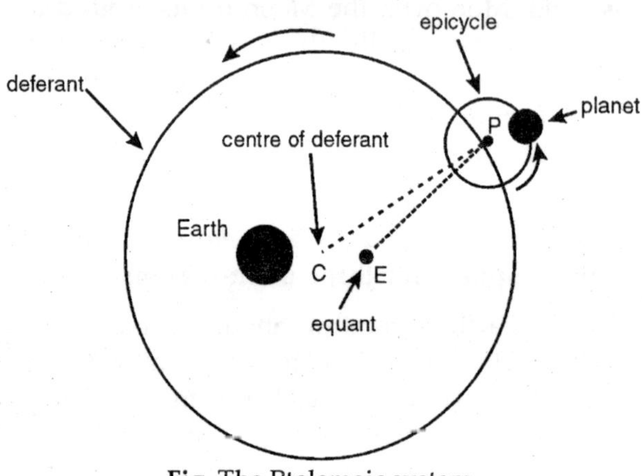

Fig. The Ptolemaic system.

First, we need to introduce some terminology. *Deferants* are large circles centred on the Earth, and *epicyles* are small circles whose centres move around the circumferences of the deferants. In the Ptolemaic system, instead of traveling around deferants, the planets move around the circumference of epicycles, which, in turn, move around the circumference of deferants. Ptolemy found, however, that this modification was insufficient to completely account for all of his data. Ptolemy's second modification to Eudoxas' model was to displace the Earth slightly from the common centre of the deferants. Moreover, Ptolemy assumed that the Sun, Moon, and planets rotate uniformly about an imaginary point, called the *equant*, which is displaced an equal distance in the opposite direction to the Earth from the centre of the deferants. In other words, Ptolemy assumed that the line *EP*, rotates uniformly, rather than the line *CP*.

Note that this diagram is not drawn to scale, and the displacement of the Earth from the centre of the deferants has been omitted for the sake of clarity. It can be seen that the Moon and the Sun do not possess epicyles. Moreover, the motions of the *inferior planets* (*i.e.*, Mercury and Venus) are closely linked to the motion of the Sun. In fact, the centres of the inferior planet epicycles move on an imaginary line connecting the Earth and the Sun. The radius vectors connecting the *superior planets* (*i.e.*, Mars, Jupiter, and Saturn) to the centres of their epicycles are always parallel to the geometric line connecting the Earth and the Sun.

Note that, in addition to the motion indicated in the diagram, all of the heavenly bodies (including the stars) rotate clockwise (assuming that we are looking down on the Earth's North pole in Figure with a period of 1 day. In fact, some planets need as many as 28 epicycles to account for all the details of their motion. These subsidiary epicycles are not shown in the diagram, for the sake of clarity.

As is quite apparent, the Ptolemaic model of the Solar System is *extremely complicated*. However, it successfully accounted for the relatively crude naked eye observations made by the ancient Greeks. The Sun-linked epicyles of the inferior planets are needed to explain why these objects always remain close to the Sun in the sky. The epicycles of the superior planets are needed to account for their occasional bouts of *retrograde motion*: *i.e.*, motion in the opposite direction to their apparent direction of rotation around the Earth. Finally, the displacement of the Earth from the centre of the deferants, as well as the introduction of the equant as the centre of uniform rotation, is needed to explain why the planets speed up slightly when they are close to the Earth (and, hence, appear brighter in the night sky), and slow down when they are further away.

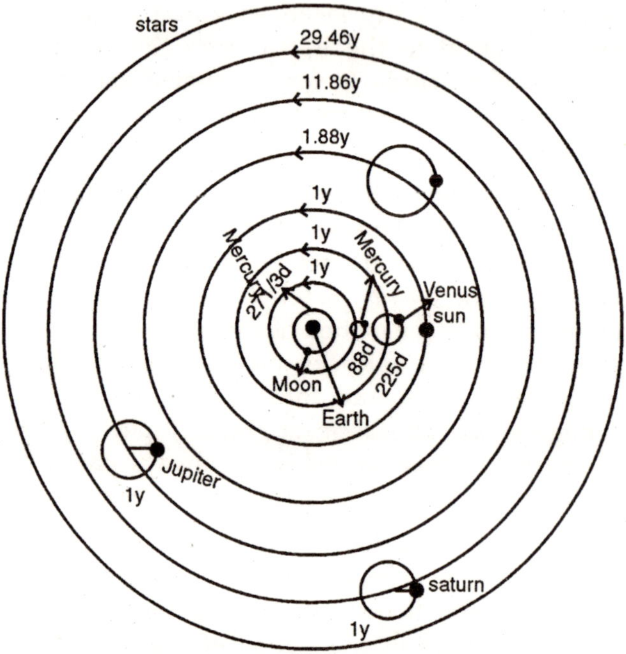

Fig. The Ptolemaic model of the Solar System.

Ptolemy's model of the Solar System was rescued from the wreck of ancient European civilization by the Roman Catholic Church, which, unfortunately, converted it into a minor article of faith, on the basis of a few references in the Bible which seemed to imply that the Earth is stationary and the Sun is moving. Consequently, this model was not subject to proper scientific criticism for over a millennium.

Having said this, few medieval or renaissance philosophers were entirely satisfied with Ptolemy's model.

Their dissatisfaction focused, not on the many epicycles (which to the modern eye seem rather absurd), but on the displacement of the Earth from the centre of the deferants, and the introduction of the equant as the centre of uniform rotation. Recall, that the only reason planetary orbits are constructed from circles in Ptolemy's model is to preserve the assumed ideal symmetry of the heavens. Unfortunately, this symmetry is severely compromised when the Earth is displaced from the apparent centre of the Universe.

This problem so perplexed the Polish priest-astronomer Nicolaus Copernicus that he eventually decided to reject the geocentric model, and revive the heliocentric model of Aristarchus.

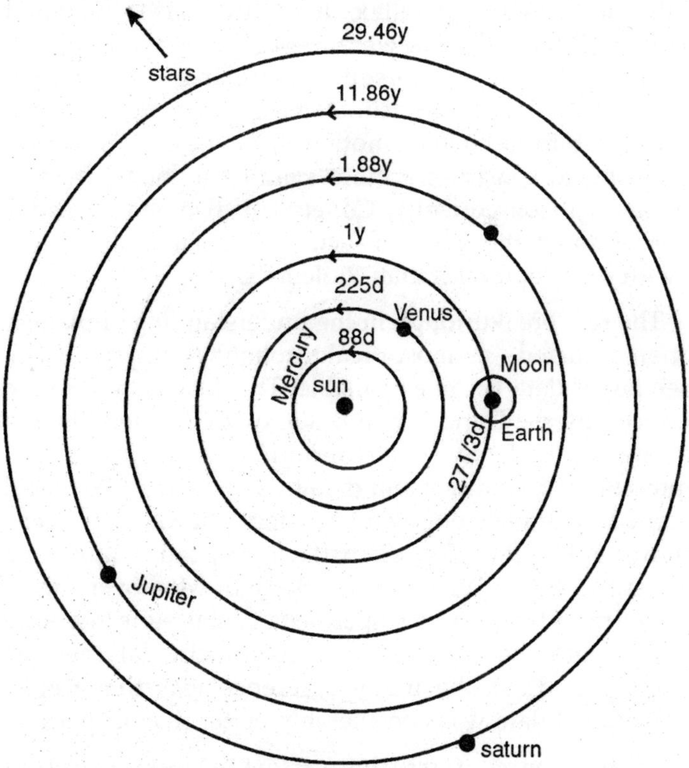

Fig. The Copernican model of the Solar System.

The planets execute uniform circular orbits about the Sun, and the Moon orbits about the Earth. Finally, the Earth revolves about its axis daily. Note that there is no displacement of the Sun from the centres of the planetary orbits, and there is no equant. Moreover, in this model, the inferior planets remain close to the Sun in the sky without any special synchronization of their orbits. The occasional retrograde motion of the superior planets has a more natural explanation than in Ptolemy's model.

Since the Earth orbits more rapidly than the superior planets, it occasionally "overtakes" them, and they appear to move backward in the night sky, in much the same manner that slow moving cars on a freeway appears to move backward to a driver overtaking them. Copernicus accounted for the lack of stellar parallax, due to the Earth's motion, by postulating that the stars were a lot further away than had previously been supposed, rendering any parallax undetectably small. Unfortunately, Copernicus insisted on retaining uniform circular motion in his model (after all, he was trying to construct a more symmetric model than that of Ptolemy). Consequently, Copernicus also had to resort to epicycles to fit the data. In fact, Copernicus' model ended up with more epicycles than Ptolemy's!

The real breakthrough in the understanding of planetary motion occurred—as most breakthroughs in physics occur—when better data became available. The data in question was produced by the Dane Tycho Brahe, who devoted his life to making naked eye astronomical observations of unprecedented accuracy and detail. This data was eventually inherited by Brahe's pupil and assistant, the German scientist Johannes Kepler Kepler fully accepted Copernicus' heliocentric theory of the Solar System. Moreover, he was just as firm a believer as Copernicus in the perfection of the heavens, and the consequent need for circular motion of planetary bodies. The main difference was that Kepler's observational data was considerably better than Copernicus'.

After years of fruitless effort, Kepler eventually concluded that no combination of circular deferants and epicycles could completely account for his data. At this stage, he started to think the unthinkable. Maybe, planetary motion was not circular after all? After more calculations, Kepler was eventually able to formulate three extraordinarily simple laws which completely accounted for Brahe's observations. These laws are as follows:

1. The planets move in elliptical orbits with the Sun at one focus.
2. A line from the Sun to any given planet sweeps out equal areas in equal time intervals.
3. The square of a planet's period is proportional to the cube of the planet's mean distance from the Sun.

Note that there are no epicyles or equants in Kepler's model of the Solar System.

Figure illustrates Kepler's second law. Here, the ellipse represents a planetary orbit, and S represents the Sun, which is located at one of the focii of the ellipse. Suppose that the planet moves from point A to point B in the same time it takes to move from point C to point D. According the Kepler's second law, the areas of the elliptic segments ASB and CSD are equal. Note that this law basically mandates that planets speed up when they move closer to the Sun.

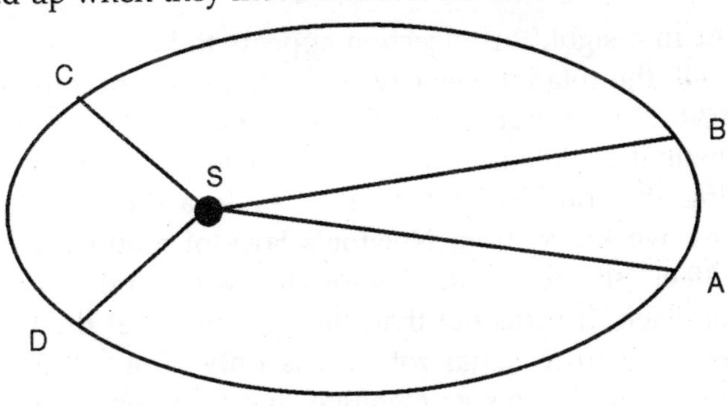

Fig. Kepler's second law.

Table illustrates Kepler's third law. The mean distance, a, and orbital period, T, as well as the ratio a^3/T^3, are listed for each of the first six planets in the Solar System. It can be seen that the ratio is indeed constant from planet to planet.

Table: Kepler's third law. Here, is the mean distance from the Sun, measured in Astronomical Units (1 AU is the mean Earth-Sun distance), and *T* is the orbital period, measured in years.

Planet	*a(AU)*	*T(yr)*	a^3/T^2
Mercury	0.387	0.241	0.998
Venus	0.723	0.615	0.999
Earth	1.000	1.000	1.000
Mars	1.524	1.881	1.000
Jupiter	5.203	11.862	1.001
Saturn	9.516	29.458	0.993

Since we have now definitely adopted a heliocentric model of the Solar System, let us discuss the ancient Greek objections to such a model, listed earlier. We have already dealt with the second objection (the absence of stellar parallax) by stating that the stars are a lot further away from the Earth than the ancient Greeks supposed. The third objection (that it is philosophically more attractive to have the Earth at the centre of the Universe) is not a valid scientific criticism. What about the first objection? If the Earth is rotating about its axis, and also orbiting the Sun, why do we not "feel" this motion?

At first sight, this objection appears to have some force. After all, the rotation velocity of the Earth's surface is about 460m/s. Moreover, the Earth's orbital velocity is approximately 30km/s. Surely, we would notice if we were moving this rapidly? Of course, this reasoning is faulty because we know, from Newton's laws of motion, that we only "feel" the acceleration associated with motion, not the motion itself. It turns out that the acceleration at the Earth's surface due to its axial rotation is only about $0.034 m/s^2$. Moreover, the Earth's acceleration due to its orbital motion is only $0.0059 m/s^2$. Nowadays, we can detect such small accelerations, but the ancient Greeks certainly could not.

Kepler correctly formulated the three laws of planetary motion in 1619. Almost seventy years later, in 1687, Isaac

Newton published his *Principia,* in which he presented, for the first time, a universal theory of motion. Newton then went on to illustrate his theory by using it to deriving Kepler's laws from first principles. Let us now discuss Newton's monumental achievement in more detail.

GRAVITY

There is one important question which we have avoided discussing until now. Why do objects fall towards the surface of the Earth? The ancient Greeks had a very simple answer to this question. According to Aristotle, all objects have a natural tendency to fall towards the centre of the Universe. Since the centre of the Earth coincides with the centre of the Universe, all objects also tend to fall towards the Earth's surface.

So, an ancient Greek might ask, why do the planets not fall towards the Earth? Well, according to Aristotle, the planets are embedded in crystal spheres which rotate with them whilst holding them in place in the firmament. Unfortunately, Ptolemy seriously undermined this explanation by shifting the Earth slightly from the centre of the Universe. However, the *coup de grace* was delivered by Copernicus, who converted the Earth into just another planet orbiting the Sun.

So, why do objects fall towards the surface of the Earth? The first person, after Aristotle, to seriously consider this question was Sir Isaac Newton. Since the Earth is not located in a special place in the Universe, Newton reasoned, objects must be attracted toward the Earth itself. Moreover, since the Earth is just another planet, objects must be attracted towards other planets as well. In fact, all objects must exert a force of attraction on all other objects in the Universe. What intrinsic property of objects causes them to exert this attractive force—which Newton termed *gravity*—on other objects? Newton decided that the crucial property was *mass.* After much thought, he was eventually able to formulate his famous law of universal gravitation:

Every particle in the Universe attracts every other particle with a force directly proportional to the product of their masses and inversely proportional to the square of the distance between them. The direction of the force is along the line joining the particles.

Incidentally, Newton adopted an inverse square law because he knew that this was the only type of force law which was consistent with Kepler's third law of planetary motion.

Consider two point objects of masses m_1 and m_2, separated by a distance r. The magnitude of the force of attraction between these objects is

$$f = G\frac{m_1 m_2}{r^2}$$

The direction of the force is along the line joining the two objects.

$$f = Gm_1m_2 / r^2$$

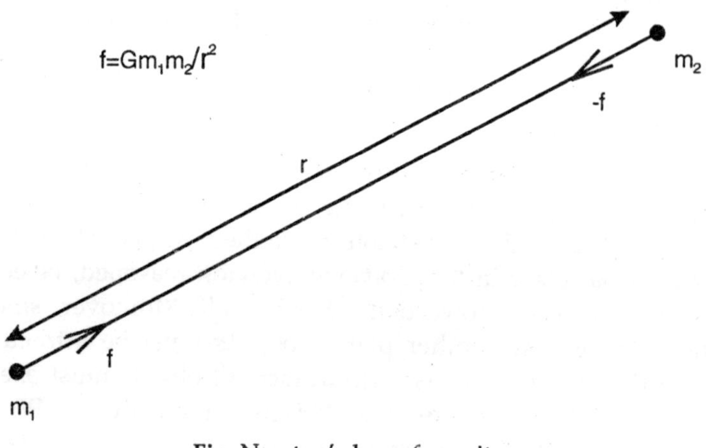

Fig. Newton's law of gravity.

Let r_1 and r_2 be the vector positions of the two objects, respectively. The vector gravitational force exerted by object 2 on object 1 can be written

$$f_{12} = G\frac{r_2 - r_1}{|r_2 - r_1|^3}$$

Likewise, the vector gravitational force exerted by object 1 on object 2 takes the form

$$f_{21} = G\frac{r_1 - r_2}{|r_1 - r_2|^3} = -f_{21}$$

The constant of proportionality, G, appearing in the above formulae is called the *gravitational constant*. Newton could only estimate the value of this quantity, which was first directly measured by Henry Cavendish in 1798. The modern value of G is

$$G = 6.6726 \times 10^{-11} \mathrm{Nm}^2/\mathrm{kg}^2.$$

Note that the gravitational constant is numerically extremely small. This implies that gravity is an intrinsically weak force. In fact, gravity usually only becomes significant if at least one of the masses involved is of astronomical dimensions (*e.g.*, it is a planet, or a star).

Let us use Newton's law of gravity to account for the Earth's surface gravity. Consider an object of mass close to the surface of the Earth, whose mass and radius are $M = 5.97 \times 10^{24}\,kg$ and $R = 6.378 \times 10^6\,m$, respectively. Newton proved, after considerable effort, that the gravitational force exerted by a spherical body (outside that body) is the same as that exerted by an equivalent point mass located at the body's centre. Hence, the gravitational force exerted by the Earth on the object in question is of magnitude

$$f = G\frac{mM}{R^2}$$

and is directed towards the centre of the Earth. It follows that the equation of motion of the object can be written

$$m\ddot{r} = -G\frac{mM}{R^2}\hat{z}$$

where $\hat{z}$ is a unit vector pointing straight upwards (*i.e.,* away from the Earth's centre). Canceling the factor m on either side of the above equation, we obtain

$$\ddot{r} = -g\ \hat{z},$$

where

$$g = \frac{Gm}{R^2}$$

$$= \frac{\left(6.673\times10^{-11}\right)\times\left(5.97\times10^{24}\right)}{\left(6.378\times10^{6}\right)^2}$$

$$= 9.79 m/s^2$$

Thus, we conclude that all objects on the Earth's surface, irrespective of 9.79 m/s^2 their mass, accelerate straight down (*i.e.,* towards the Earth's centre) with a constant acceleration of. This estimate for the acceleration due to gravity is slightly off the conventional value of 9.81 m/s^2 because the Earth is actually not quite spherical.

Since Newton's law of gravitation is universal, we immediately conclude that any spherical body of mass M and radius R possesses a surface gravity g given by the following formula:

$$\frac{g}{g} = \frac{M\ /\text{-}M}{(R\ /\ R\ )^2}.$$

Table shows the surface gravity of various bodies in the Solar System, estimated using the above expression. It can be seen that the surface gravity of the Moon is only about one fifth of that of the Earth. No wonder Apollo astronauts were able to jump so far on the Moon's surface! Prospective Mars colonists should note that they will only weigh about a third of their terrestrial weight on Mars.

Table: The mass, M, radius, R, and surface gravity, g, of various bodies in the Solar System. All quantities are expressed as fractions of the corresponding terrestrial quantity.

Body	$M/M_{\oplus}$	$R/R_{\oplus}$	$g/g_{\oplus}$
Sun	3.33×10^5	109.0	28.1
Moon	0.0123	0.273	0.17
Mercury	0.0553	0.383	0.38
Venus	0.816	0.949	0.91
Earth	1.000	1.000	1.000
Mars	0.108	0.533	0.38
Jupiter	318.3	11.21	2.5
Saturn	95.14	9.45	1.07

GRAVITATIONAL POTENTIAL ENERGY

The gravity is a conservative force, and, therefore, has an associated potential energy. Let us obtain a general formula for this energy. Consider a point object of mass m, which is a radial distance from another point object of mass M. The gravitational force acting on the first mass is of magnitude $f = GM/r^2$, and is directed towards the second mass. Imagine that the first mass moves radially away from the second mass, until it reaches infinity. What is the change in the potential energy of the first mass associated with this shift? According to Equation,

$$U(\infty)-U(r) = -\int_r^{\infty}[-f(r)]\,dr$$

There is a minus sign in front of f because this force is oppositely directed to the motion. The above expression can be integrated to give

$$U(r) = -\frac{GMm}{r}$$

Here, we have adopted the convenient normalization that the potential energy at infinity is zero. According to the above formula, the gravitational potential energy of a mass m located a distance r from a mass M is simply $-GMm/r$.

Consider an object of mass z moving close to the Earth's surface. The potential energy of such an object can be written

$$U = -\frac{GM\ m}{R\ +z}$$

where M and R are the mass and radius of the Earth, respectively, and z is the vertical height of the object above the Earth's surface. In the limit that $z \ll R$, the above expression can be expanded using the binomial theorem to give $U \simeq \frac{GM\ m}{R} + \frac{GM\ m}{R^2} z,$

Since potential energy is undetermined to an arbitrary additive constant, we could just as well write $U \simeq mgz$,

where $g = GM\ \ / R^2$ is the acceleration due to gravity at the Earth's surface. Of course, the above formula is equivalent to the formula derived earlier on in this course.

For an object of mass m and speed υ, moving in the gravitational field of a fixed object of mass M, we expect the total energy,

$$E = K + U,$$

to be a constant of the motion. Here, the kinetic energy is written $K = (1/2)m\upsilon^2$, whereas the potential energy takes the form $U = -GMm/r$. Of course, r is the distance between the two objects. Suppose that the fixed object is a sphere of radius R. Suppose, further, that the second object is launched from the surface of this sphere with some velocity υ_{esc} which is such that it *only just* escapes the sphere's gravitational influence.

After the object has escaped, it is a long way away from the sphere, and hence $U = 0$. Moreover, if the object only just escaped, then we also expect , since the object will have expended all of its initial kinetic energy escaping from the sphere's gravitational well. We conclude that our object possesses zero net energy: *i.e.*, $E = K+U = 0$. Since E is a constant of the motion, it follows that at the launch point

$$E = \frac{1}{2}m\upsilon_{esc}^2 - \frac{GMm}{R}$$
$$= 0$$

This expression can be rearranged to give

$$\upsilon_{esc} = \sqrt{\frac{2GM}{R}}$$

The quantity υ_{esc} is known as the *escape velocity*. Objects launched from the surface of the sphere with velocities exceeding this value will eventually escape from the sphere's gravitational influence. Otherwise, the objects will remain in orbit around the sphere, and may eventually strike its surface. Note that the escape velocity is independent of the object's mass and launch direction (assuming that it is not straight into the sphere).

The escape velocity for the Earth is

$$\upsilon_{esc} = \sqrt{\frac{2GM}{R}}$$
$$= \sqrt{\frac{2\times(6.673\times10^{-11})\times(5.97\times10^{24})}{6.378\times10^{6}}}$$
$$= 11.2 \text{ km/s}$$

Clearly, NASA must launch deep space probes from the surface of the Earth with velocities which exceed this value if they are to have any hope of eventually reaching their targets.

SATELLITE ORBITS

Consider an artificial satellite executing a circular orbit of radius r around the Earth. Let ω be the satellite's orbital angular velocity. The satellite experiences an acceleration towards the Earth's centre of magnitude $\omega^2 r$. Of course, this acceleration is provided by the gravitational attraction between the satellite and the Earth, which yields an acceleration of magnitude GM / r^2. It follows that

$$\omega^2 r = \frac{GM}{r^2}$$

Suppose that the satellite's orbit lies in the Earth's equatorial plane. Moreover, suppose that the satellite's orbital angular velocity just matches the Earth's angular velocity of rotation. In this case, the satellite will appear to hover in the *same place* in the sky to a stationary observer on the Earth's surface. A satellite with this singular property is known as a *geostationary satellite.*

Virtually all of the satellites used to monitor the Earth's weather patterns are geostationary in nature. Communications satellites also tend to be geostationary. Of course, the satellites which beam satellite-TV to homes across the world *must* be geostationary—otherwise, you would need to install an expensive tracking antenna on top of your house in order to pick up the transmissions. Incidentally, the person who first envisaged rapid global telecommunication via a network of geostationary satellites was the science fiction writer Arthur C. Clarke in 1945.

Let us calculate the orbital radius of a geostationary satellite. The angular velocity of the Earth's rotation is

$$\omega = \frac{2\pi}{24 \times 60 \times 60}$$

$$= 7.27 \times 10^{-5} \text{rad./s.}$$

It follows from Equation, that

$$r_{geo} = \left(\frac{GM}{\omega^2}\right)$$

$$= \left(\frac{\left(6.673\times10^{-11}\right)\times\left(5.97\times10^{24}\right)}{\left(7.27\times10^{-5}\right)^2}\right)^{1/3}$$

$$= 4.22\times10^{7}\text{m}$$

$$= 6.62R$$

Thus, a geostationary satellite must be placed in a circular orbit whose radius is *exactly* 6.62 times the Earth's radius.

PLANETARY ORBITS

Let us now see whether we can use Newton's universal laws of motion to derive Kepler's laws of planetary motion. Consider a planet orbiting around the Sun. It is convenient to specify the planet's instantaneous position, with respect to the Sun, in terms of the *polar coordinates* r and θ. As illustrated in figure, r is the radial distance between the planet and the Sun, whereas θ is the angular bearing of the planet, from the Sun, measured with respect to some arbitrarily chosen direction.

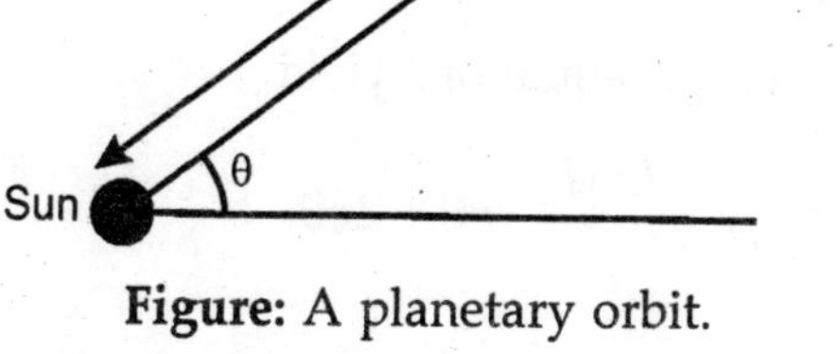

Figure: A planetary orbit.

Let us define two unit vectors, e_r and e_θ. (A unit vector is simply a vector whose length is unity.) The *radial* unit vector e_r always points from the Sun towards the instantaneous position of the planet. Moreover, the *tangential* unit vector e_θ is always normal to e_r, in the direction of increasing θ. We demonstrated that when acceleration is written in terms of polar coordinates, it takes the form

$$a = a_r e_r + a_\theta e_\theta$$

where

$$a_r = \ddot{r} - r\dot{\theta}^2$$

$$a_\theta = \ddot{r}\theta - 2\dot{r}\dot{\theta}$$

These expressions are more complicated that the corresponding cartesian expressions because the unit vectors e_r and e_θ *change direction* as the planet changes position.

Now, the planet is subject to a single force: *i.e.*, the force of gravitational attraction exerted by the Sun. In polar coordinates, this force takes a particularly simple form (which is why we are using polar coordinates):

$$f = -\frac{GM\ m}{r^2} e_r$$

The minus sign indicates that the force is directed towards, rather than away from, the Sun.

According to Newton's second law, the planet's equation of motion is written

$$ma = f$$

The above four equations yield

$$\ddot{r} - r\dot{\theta}^2 = -\frac{GM}{r^2}, r\ddot{\theta} + 2\dot{r}\dot{\theta}$$

Equation reduces to $\frac{d}{dt}\left(r^2\dot{\theta}\right) = 0$, or $r^2\dot{\theta} = h$ where h is a *constant of the motion*. What is the physical interpretation of

h? That the angular momentum vector of a point particle can be written $i = mr \times v$ For the case in hand, $r = r\,e_r$ and $v = \dot{r}e_r + r\dot{\theta}e_\theta$. Hence, $l = mr\upsilon_\theta = mr^2\dot{\theta}$, yielding $h = \frac{l}{m}$. Clearly, h represents the *angular momentum* (per unit mass) of our planet around the Sun. Angular momentum is conserved (*i.e.*, h is constant) because the force of gravitational attraction between the planet and the Sun exerts *zero torque* on the planet. that torque is the rate of change of angular momentum.The torque is zero because the gravitational force is *radial* in nature: *i.e.*, its line of action passes through the Sun, and so its associated lever arm is of length zero.

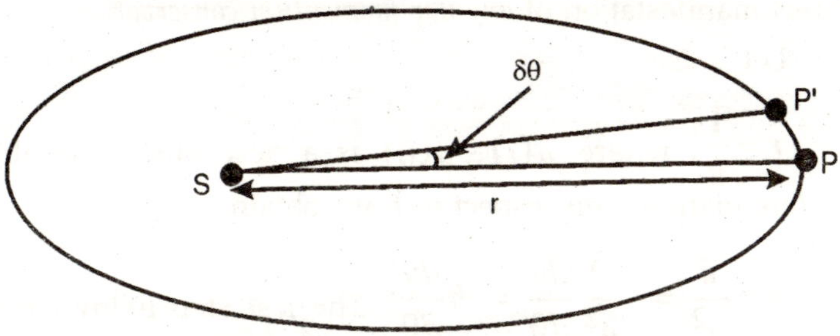

Figure: The origin of Kepler's second law.

The quantity h has another physical interpretation. Consider figure, Suppose that our planet moves from P to P' in the short time interval δt. Here, S represents the position of the Sun. The lines SP and SP' are both approximately of length. Moreover, using simple trigonometry, the line PP' is of length $r\delta\theta$, where $\delta\theta$ is the small angle through which the line joining the Sun and the planet rotates in the time interval δt. The area of the triangle PSP' is approximately

$$\delta A = \frac{1}{2} \times r\delta\theta \times r:$$

i.e., half its base times its height. Of course, this area represents the area swept out by the line joining the Sun and the planet in the time interval δt. Hence, the rate at which this area is swept is given by

$$\lim_{\delta t \to 0} \frac{\delta A}{\delta t} = \frac{1}{2} r^2 \lim_{\delta t \to 0} \frac{\delta \theta}{\delta t}$$

$$= \frac{r^2 \theta}{2}$$

$$= \frac{h}{2}$$

Clearly, the fact that h is a constant of the motion implies that the line joining the planet and the Sun sweeps out area at a *constant rate*: *i.e.*, the line sweeps equal areas in equal time intervals. But, this is just Kepler's second law. We conclude that Kepler's second law of planetary motion is a direct manifestation of *angular momentum conservation*.

Let

$r = \frac{1}{u}$, where $u(t) = u(\theta)$ is a new radial variable. Differentiating with respect to t, we obtain

$\dot{r} = \frac{\dot{u}}{u^2} = -\frac{\dot{\theta}}{u^2} \frac{du}{d\theta} = -h \frac{du}{d\theta}$. The last step follows from the fact that $\dot{\theta} = hu^2$.

Differentiating a second time with respect to t, we obtain

$$\ddot{r} = -h \frac{d}{dt}\left(\frac{du}{d\theta}\right) = -h\dot{\theta} \frac{d^2 u}{d\theta^2} = -h^2 u^2 \frac{d^2 u}{d\theta^2}$$

Equations can be combined to give $\frac{d^2 u}{d\theta^2} + u = \frac{GM}{h^2}$

This equation possesses the fairly obvious general solution

$$u = A\cos(\theta - \theta_0) + \frac{GM}{2}$$

where A and θ_0 are arbitrary constants. The above formula can be inverted to give the following simple orbit equation for our planet: $r = \frac{1}{A\cos(\theta = \theta_0) + GM / h^2}$ The constant θ_0 merely determines the orientation of the orbit.

Since we are only interested in the orbit's *shape*, we can set this quantity to zero without loss of generality. Hence, our orbit equation reduces to $r = r_0 \frac{1+e}{1+e\cos\theta}$, where $e = \frac{Ah^2}{GM}$, and $r_0 = \frac{h^2}{GM(1+e)}$, Formula is the standard equation of an *ellipse* (assuming), with the origin at a focus. Hence, we have now proved Kepler's first law of planetary motion. It is clear that r_0 is the radial distance at $\theta = 0$. The radial distance at $\theta = \pi$ is written

$$r_1 = r_0 \frac{1+e}{1-e}$$

Here, r_0 is termed the *perihelion* distance (*i.e.*, the closest distance to the Sun) and r_1 is termed the *aphelion* distance (*i.e.*, the furthest distance from the Sun). The quantity $e = \frac{r_1 - r_0}{r_1 + r_0}$ is termed the *eccentricity* of the orbit, and is a measure of its departure from circularity. Thus, e = 0corresponds to a purely circular orbit, whereas corresponds to a highly elongated orbit. As specified in Table, the orbital eccentricities of all of the planets (except Mercury) are fairly small.

Table: The orbital eccentricities of various planets in the Solar System.

Planet	e
Mercury	0.206
Venus	0.007
Earth	0.017
Mars	0.093
Jupiter	0.048
Saturn	0.056

According to equation, a line joining the Sun and an orbiting planet sweeps area at the constant rate $h/2$. Let T be the planet's orbital period. We expect the line to sweep out the *whole area* of the ellipse enclosed by the planet's orbit in the time interval T. Since the area of an ellipse is πab, where a and b are the *semi-major* and *semi-minor* axes, we can write

$$T = \frac{\pi ab}{h/2}$$

The relationship between the aphelion distance, the perihelion distance, and the semi-major and semi-minor axes of a planetary orbit. It is clear, from the figure, that the semi-major axis is just the mean of the aphelion and perihelion distances: *i.e.*,

$$a = \frac{r_0 + r_1}{2}$$

Thus, a is essentially the planet's mean distance from the Sun. Finally, the relationship between a, b, and the eccentricity, e, is given by the well-known formula

$$\frac{b}{a} = \sqrt{1-e^2}$$

This formula can easily be obtained from Equation.

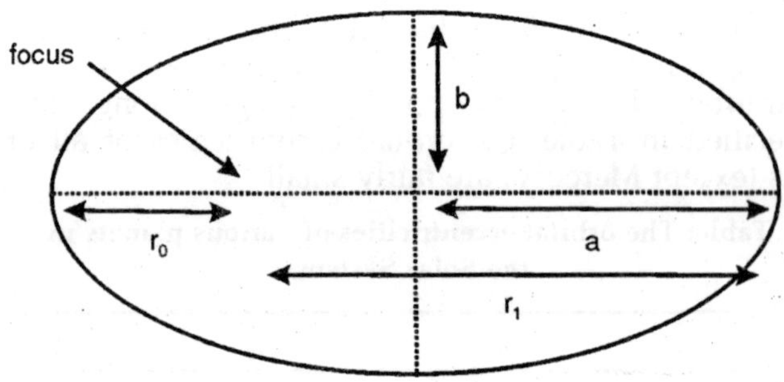

Fig. Anatomy of a planetary orbit.

Equations can be combined to give

$$a = \frac{h^2}{2Gm}\left(\frac{1}{1+e} + \frac{1}{1-e}\right)$$

$$= \frac{h^2}{GM\left(1-e^2\right)}$$

It follows, from Equations. that the orbital period can be written $T = \frac{2\pi}{\sqrt{GM}} a^{3/2}$. Thus, the orbital period of a planet is proportional to its mean distance from the Sun to the power 3/2 —the constant of proportionality being the same for all planets. Of course, this is just Kepler's third law of planetary motion.

Problem: Callisto is the eighth of Jupiter's moons: its mass and radius are $M = 1.08 \times 10^{23}\, kg$ and, respectively. What is the gravitational acceleration on the surface of this moon?

Soluiton: The surface gravitational acceleration on a spherical body of mass M and radius R is simply

$$g = \frac{GM}{R^2}$$

Hence,

$$g = \frac{\left(6.673\times10^{-11}\right)\times\left(1.08\times10^{23}\right)}{\left(2.403\times10^{6}\right)^2}$$

$$= 1.25 \text{ m/s}^2.$$

Problem: A rocket is located a distance 3.5 times the radius of the Earth above the Earth's surface. What is the rocket's free-fall acceleration?

Solution: Let R be the Earth's radius. The distance of the rocket from the centre of the Earth is $r_1 = (3.5+1)R = 4.5R$. We know that the free-fall acceleration of the rocket when its distance from the Earth's centre is $r_0 = R$ (*i.e.,* when it is at the Earth's surface) is $g_0 = 9.81 m/s^2$. Moreover, we know that gravity is an inverse-square law (*i.e.*, $g\alpha 1/r^2$). Hence, the rocket's acceleration is

$$g_1 = g_0\left(\frac{r_0}{r_1}\right)^2$$

$$= \frac{9.81\times 1}{(4.5)^2}$$

$$= 0.484 m/s^2$$

Problem: A satellite moves in a circular orbit around the Earth with speed $\upsilon = 6000 m/s$. Determine the satellite's altitude above the Earth's surface. Determine the period of the satellite's orbit. The Earth's mass and radius are $M = 5.97\times 10^{24} kg$ and, $R = 6.378\times 10^6 m$ respectively.

Solution: The acceleration of the satellite towards the centre of the Earth is υ^2/r, where is its orbital radius. This acceleration must be provided by the acceleration GM/r^2 due to the Earth's gravitational attraction. Hence,

$$\frac{\upsilon^2}{r} = \frac{GM}{r^2}$$

The above expression can be rearranged to give

$$r = \frac{GM}{r^2}$$

$$= \frac{(6.673\times 10^{-11})\times(5.97\times 10^{24})}{(6000)^2}$$

$$= 1.107\times 10^7 m.$$

Thus, the satellite's altitude above the Earth's surface is

$$h = r - R$$

$$= 1.107\times10^{17} - 6.378\times10^{6}$$

$$= 4.69\times10^{6} m \cdot$$

The satellite's orbital period is simply

$$T = \frac{2\pi r}{\upsilon}$$

$$= \frac{2\times\pi\times\left(1.107\times10^{7}\right)}{6000}$$

$$= 3.22 \text{ hours.}$$

Problem: The distance of closest approach of Halley's comet to the Sun is $0.57AU$. (1 AU is the mean Earth-Sun distance.) The greatest distance of the comet from the Sun is 35AU. The comet's speed at closest approach is $54km/s$. What is its speed when it is furthest from the Sun?

Solution: At perihelion and aphelion, the comet's velocity is perpendicular to its position vector from the Sun. Hence, at these two special points, the comet's angular momentum (around the Sun) takes the particularly simple form

$$l = mru$$

Here, *m* is the comet's mass, *r* is its distance from the Sun, and *u* is its speed. According to Kepler's second law, the comet orbits the Sun with *constant* angular momentum. Hence, we can write

$$r_0 u_0 = r_1 u_1$$

where r_0 and u_0 are the perihelion distance and speed, respectively, and r_1 and u_1 are the corresponding quantities at aphelion. We are told that $r_0 = 0.57AU$, $r_1 = 35AU$, and $u_0 = 54km/s$. It follows that

$$u_1 = \frac{u_0 r_0}{r_1}$$

$$= \frac{54 \times 0.57}{35}$$

$$= 0.879 km/s$$

Problem: A planet is in circular orbit around a star. The period and radius of the orbit are $T = 43 \times 10^7$ and $r = 2.34 \times 10^{11} m$, respectively. Calculate the mass of the star.

Solution: Let ω be the planet's orbital angular velocity. The planet accelerates towards the star with acceleration $\omega^2 r$. The acceleration due to the star's gravitational attraction is GM / r^2, where M is the mass of the star. Equating these accelerations, we obtain

$$\omega^2 r = \frac{GM}{rh2}$$

Now,

$$T = \frac{2\pi}{\omega}$$

Hence, combining the previous two expressions, we get

$$M_* = \frac{4\pi^2 r^3}{GT^2}$$

Thus, the mass of the star is

$$M = \frac{4 \times \pi^2 \left(2.34 \times 10^{11}\right)^3}{\left(6.673 \times 10^{-11}\right) \times \left(4.3 \times 10^7\right)^2}$$

$$= 4.01 \times 10^{30} kg$$

Problem: What is the minimum energy required to launch a probe of mass $m = 120 kg$ into outer space? The Earth's mass and radius are $M = 5.97 \times 10^{24} kg$ and $R = 6.378 \times 10^6 m$, respectively.

Solution: The energy which must be given to the probe should just match the probe's gain in potential energy as it travels from the Earth's surface to outer space. By definition, the probe's potential energy in outer space is zero. The potential energy of the probe at the Earth's surface is

$$U = -\frac{GM}{R}$$

$$= \frac{(6.673\times10^{-11})\times(5.97\times10^{24})}{(6.378\times10^{6})}$$

$$= 7.495\times10^{9}\,J$$

Thus, the gain in potential energy, which is the same as the minimum launch energy, is $7.495\times10^{9}\,J$.

Chapter 5

Angular Momentum

ANGULAR MOMENTUM OF A POINT PARTICLE

Consider a particle of mass m, position vector r, and instantaneous velocity v, which rotates about an axis passing through the origin of our coordinate system. We know that the particle's linear momentum is written

$$P = mv,$$

and satisfies

$$\frac{dp}{dt} = f,$$

where f is the force acting on the particle. Let us search for the rotational equivalent of p.

Consider the quantity

$$l = r \times p.$$

This quantity—which is known as *angular momentum*—is a vector of magnitude

$$l = r\,p\,\sin\theta$$

where θ is the angle subtended between the directions of r and P. The direction of is defined to be mutually perpendicular to the directions of r and P, in the sense given by the right-hand grip rule. In other words, if vector r rotates onto vector P (through an angle less than 180°), and the fingers of the right-hand are aligned with this rotation, then the thumb of the right-hand indicates the direction of l.

$$l = rp\sin\theta$$

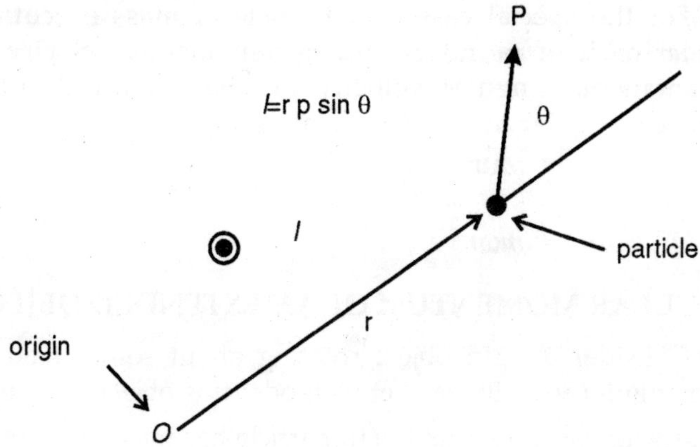

Fig. Angular momentum of a point particle about the origin.

Let us differentiate equation with respect to time. We obtain

$$\frac{dl}{dt} = \dot{r} \times p \times r \times \dot{p}$$

Note that the derivative of a vector product is formed in much the same manner as the derivative of an ordinary product, except that the order of the various terms is preserved. Now, we know that $\dot{r} = v = p/m$ and $\dot{p} = f$. Hence, we obtain

$$\frac{dl}{dt} = \frac{p \times p}{m} + r \times f.$$

However, $p \times p = 0$, since the vector product of two parallel vectors is zero. Also,

$$r \times f = \tau$$

where τ is the torque acting on the particle about an axis passing through the origin. We conclude that

$$\frac{dl}{dt} = \tau$$

Of course, this equation is analogous to equation which suggests that angular momentum, *l*, plays the role of linear momentum, *p*, in rotational dynamics.

For the special case of a particle of mass executing a *circular* orbit of radius r, with instantaneous velocity and instantaneous angular velocity ω, the magnitude of the particle's angular momentum is simply

$$l = m\upsilon r$$

$$= m\omega r^2.$$

ANGULAR MOMENTUM OF AN EXTENDED OBJECT

Consider a rigid object rotating about some fixed axis with angular velocity ω. Let us model this object as a swarm of particles. Suppose that the th particle has mass m_i, position vector r_i, and velocity v_i. Incidentally, it is assumed that the object's axis of rotation passes through the origin of our coordinate system. The total angular momentum of the object, L, is simply the vector sum of the angular momenta of the N particles from which it is made up. Hence,

$$L = \sum_{i=1,N} m_i r_i \times v_i.$$

Now, for a rigidly rotating object we can write

$$v_i = \omega \times r_i$$

Let

$$\omega = \omega k,$$

where k is a unit vector pointing along the object's axis of rotation (in the sense given by the right-hand grip rule). It follows that

$$L = \omega \sum_{i=1,N} m_i r_i \times (k \times r_i)$$

Let us calculate the component of L along the object's rotation axis—*i.e.*, the component along the k axis. We can write

$$L_k = L \cdot k$$

$$= \omega \sum_{i=1,N} m_i k \cdot r_i \times (k \times r_i).$$

However, since $a \cdot b \times c = a \times b \cdot c$, the expression can be rewritten

$$L_k = \omega \sum_{i=1,N} m_i (k \times r_i) \cdot (k \times r_i)$$

$$= \omega \sum_{i=1,N} m_i |k \times r_i|^2$$

Now,

$$\sum_{i=1,N} m_i |k \times r_i|^2 = I_k,$$

where I_k is the moment of inertia of the object about the k axis. Hence, it follows that

$$L_k = I_k \omega$$

According to the formula, the component of a rigid body's angular momentum vector along its axis of rotation is simply the product of the body's moment of inertia about this axis and the body's angular velocity. Does this result imply that we can automatically write

$$L = I\omega?$$

Unfortunately, in general, the answer to the above question is *no*! This conclusion follows because the body may possess non-zero angular momentum components about axes perpendicular to its axis of rotation. Thus, in general, the angular momentum vector of a rotating body is *not* parallel to its angular velocity vector. This is a major difference from translational motion, where linear momentum is always found to be parallel to linear velocity.

For a rigid object rotating with angular velocity $\omega = (\omega_x, \omega_y, \omega_z)$, we can write the object's angular momentum $L = (L_x, L_y, L_z)$ in the form

$$L_x = I_x \omega_x,$$

$$L_y = I_y \omega_y,$$

$$L_z = I_z\omega_z$$

where I_x is the moment of inertia of the object about the -axis, *etc.* Here, it is again assumed that the origin of our coordinate system lies on the object's axis of rotation. Note that the equations are only valid when the *x*-, y-, and -axes are aligned in a certain very special manner—in fact, they must be aligned along the so-called *principal axes* of the object (these axes invariably coincide with the object's main symmetry axes). Note that it is always possible to find three, mutually perpendicular, principal axes of rotation which pass through a given point in a rigid body. Reconstructing *L* from its components, we obtain

$$L = \left(L_x\omega_x\hat{x} + L_y\omega_y\hat{y} + L_z\omega_z\hat{z}\right)$$

where $\hat{x}$ is a unit vector pointing along the *x*-axis, *etc.* It is clear, from the equation, that the reason *L* is not generally parallel to ω is because the moments of inertia of a rigid object about its different possible axes of rotation are *not generally the same.* In other words, if $I_x = I_y = I_z = I$ then $L = I\omega$, and the angular momentum and angular velocity vectors are always parallel. However, if $I_x \neq I_y \neq I_z$, which is usually the case, then *L* is not, in general, parallel to ω.

Although equation suggests that the angular momentum of a rigid object is not generally parallel to its angular velocity, this equation also implies that there are, at least, *three* special axes of rotation for which this is the case. Suppose, for instance, that the object rotates about the *z*-axis, so that $\omega = \omega_z\hat{z}$. It follows from equation that

$$L = I_z\omega_z\hat{z} = I_z\omega$$

Thus, in this case, the angular momentum vector is parallel to the angular velocity vector. The same can be said for rotation about the *x*- or *y*- axes. We conclude that when a rigid object rotates about one of its principal axes then its angular momentum is parallel to its angular velocity, but not, in general, otherwise.

How can we identify a principal axis of a rigid object? At the simplest level, a principal axis is one about which the object possesses *axial symmetry*. Assuming that the object can be modeled as a swarm of particles—for every particle of mass *m*, located a distance from the origin, and subtending an θ angle with the rotation axis, there must be an identical particle located on diagrammatically the opposite side of the rotation axis.

As shown in the diagram, the angular momentum vectors of such a matched pair of particles can be added together to form a resultant angular momentum vector which is *parallel* to the axis of rotation. Thus, if the object is composed entirely of matched particle pairs then its angular momentum vector must be parallel to its angular velocity vector. The generalization of this argument to deal with continuous objects is fairly straightforward. For instance, symmetry implies that any axis of rotation which passes through the centre of a uniform sphere is a principal axis of that object. Likewise, a perpendicular axis which passes through the centre of a uniform disk is a principal axis. Finally, a perpendicular axis which passes through the centre of a uniform rod is a principal axis.

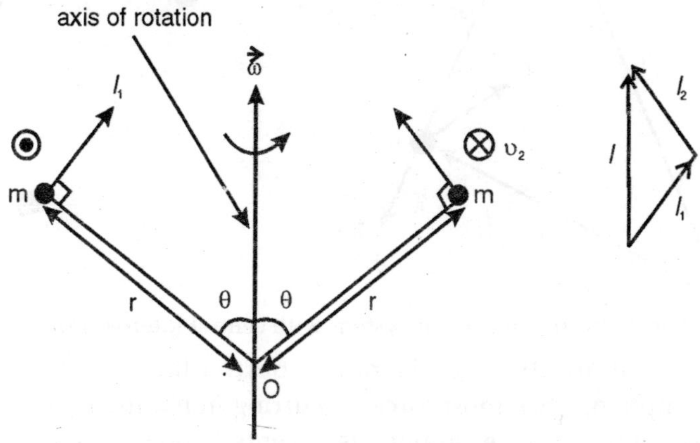

Fig. A principal axis of rotation.

ANGULAR MOMENTUM OF A MULTI-COMPONENT SYSTEM

Consider a system consisting of N mutually interacting point particles. Such a system might represent a true multi-component system, such as an asteroid cloud, or it might represent an extended body. Let the ith particle, whose mass is m_i, be located at vector displacement r_i. Suppose that this particle exerts a force f_{ji} on the jth particle. By Newton's third law of motion, the force f_{ij} exerted by the jth particle on the ith is given by

$$f_{ij} = -f_{ji} .$$

Let us assume that the internal forces acting within the system are *central forces—i.e.*, the force f_{ij}, acting between particles i and j, is directed along the *line of centres* of these particles. In other words, $f_{ij} \alpha \left(r_i - r_j\right)$.

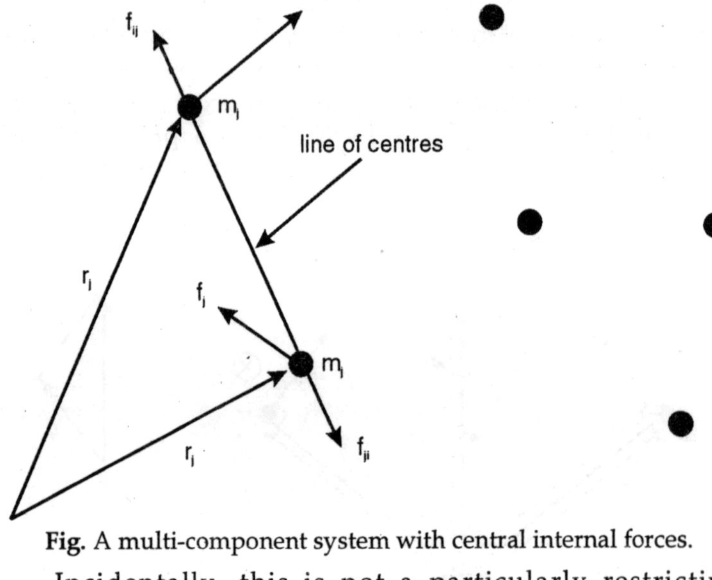

Fig. A multi-component system with central internal forces.

Incidentally, this is not a particularly restrictive assumption, since most forces occurring in nature are central forces. For instance, gravity is a central force, electrostatic forces are central, and the internal stresses acting within a rigid body are approximately central. Suppose, finally, that the th particle is subject to an external force *Fi*.

The equation of motion of the *i*th particle can be written

$$\dot{p}_i = \sum_{j=1,N}^{j\neq i} f_{ij} + F_i$$

Taking the vector product of this equation with the position vector r_i, we obtain

$$r_i \times \dot{p}_i = \sum_{j=1,N}^{j\neq i} r_i \times f_{ij} + r_i \times F_i$$

Now, we have already seen that

$$r_i \times \dot{p}_i = \frac{d(r_i \times p_i)}{dt}.$$

We also know that the total angular momentum, *L*, of the system (about the origin) can be written in the form

$$L = \sum_{i=1,N} r_i \times p_i.$$

Hence, summing equation over all particles, we obtain

$$\frac{dl}{dt} = \sum_{i,j=i,N}^{i\neq j} r_i \times f_{ij} + \sum_{i=1,N} r_i \times F_i.$$

Consider the first expression on the right-hand side of equation, A general term, $r_i \times f_{ij}$, in this sum can always be paired with a matching term, $r_i \times f_{ij}$, in which the indices have been swapped. Making use of Equation, the sum of a general matched pair can be written

$$r_i \times f_{ij} + r_j \times f_{ij} = (r_i - r_j) \times f_{ij}$$

However, if the internal forces are central in nature then f_{ij} is parallel to $(r_i - r_j)$. Hence, the vector product of these two vectors is zero. We conclude that $r_i \times f_{ij} + r_j \times f_{ij} = 0$

for any values of j and . Thus, the first expression on the right-hand side of Equation. sums to zero. We are left with

$$\frac{dL}{dt} = \tau$$

where

$$\tau = \sum_{i=1,N} r_i \times F_i$$

is the net external torque acting on the system (about an axis passing through the origin). Of course, equation is simply the rotational equation of motion for the system taken as a whole.

Suppose that the system is isolated, such that it is subject to zero net external torque. It follows from equation that, in this case, the total angular momentum of the system is a conserved quantity. To be more exact, the components of the total angular momentum taken about any three independent axes are individually conserved quantities. Conservation of angular momentum is an extremely useful concept which greatly simplifies the analysis of a wide range of rotating systems. Let us consider some examples.

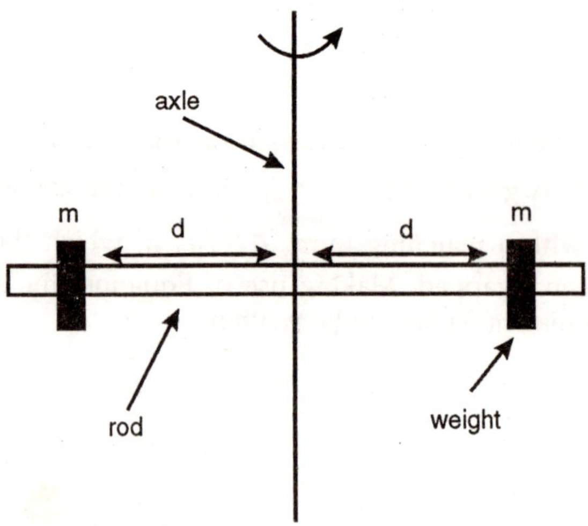

Fig. Two movable weights on a rotating rod.

Suppose that two identical weights of mass m are attached to a light rigid rod which rotates without friction about a perpendicular axis passing through its mid-point. Imagine that the two weights are equipped with small motors which allow them to travel along the rod: the motors are synchronized in such a manner that the distance of the two weights from the axis of rotation is always the same. Let us call this common distance d, and let ω be the angular velocity of the rod. How does the angular velocity ω change as the distance d is varied?

Note that there are no external torques acting on the system. It follows that the system's angular momentum must remain constant as the weights move along the rod. Neglecting the contribution of the rod, the moment of inertia of the system is written

$$I = 2md^2$$

Since the system is rotating about a principal axis, its angular momentum takes the form

$$L = I\omega = 2md^2\omega .$$

If L is a constant of the motion then we obtain

$$\omega d^2 = \textit{constant}.$$

In other words, the system spins *faster* as the weights move *inwards* towards the axis of rotation, and *vice versa*. This effect is familiar from figure skating. When a skater spins about a vertical axis, her angular momentum is approximately a conserved quantity, since the ice exerts very little torque on her. Thus, if the skater starts spinning with outstretched arms, and then draws her arms inwards, then her rate of rotation will spontaneously increase in order to conserve angular momentum. The skater can slow her rate of rotation by simply pushing her arms outwards again.

Suppose that a bullet of mass m and velocity υ strikes, and becomes embedded in, a stationary rod of mass M and length $2b$ which pivots about a frictionless perpendicular axle passing through its mid-point. Let the bullet strike the rod normally a distance d from its axis of rotation.What is the instantaneous angular velocity of the rod (and bullet) immediately after the collision?

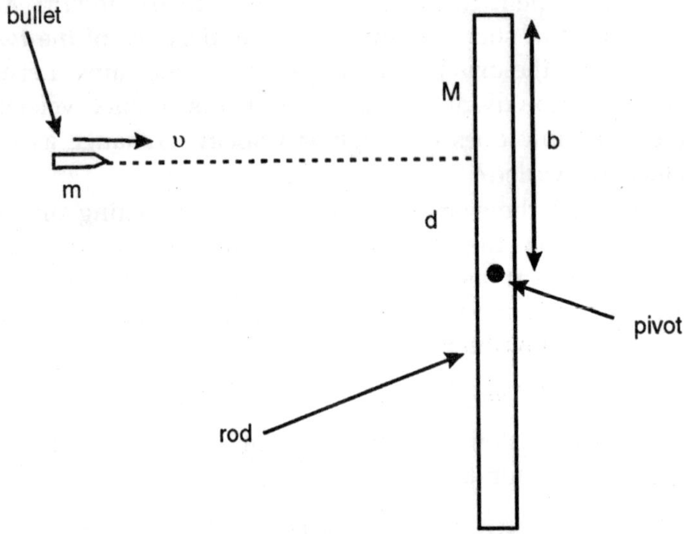

Figure: A bullet strikes a pivoted rod.

Taking the bullet and the rod as a whole, this is again a system upon which no external torque acts. Thus, we expect the system's net angular momentum to be the same before and after the collision. Before the collision, only the bullet possesses angular momentum, since the rod is at rest. As is easily demonstrated, the bullet's angular momentum about the pivot point is

$$l = m\upsilon d:$$

i.e., the product of its mass, its velocity, and its distance of closest approach to the point about which the angular momentum is measured—this is a general result (for a point particle). After the collision, the bullet lodges a distance from the pivot, and is forced to co-rotate with the rod. Hence, the angular momentum of the bullet after the collision is given by

$$l' = md^2\omega$$

where ω is the angular velocity of the rod. The angular momentum of the rod after the collision is

$$L = I\omega$$

where $I = (1/12)M(2b)^2 = (1/3)Mb^2$ is the rod's moment of inertia (about a perpendicular axis passing through its mid-point). Conservation of angular momentum yields

$$l = l' + L$$

or

$$\omega = \frac{m\upsilon d}{1 + md^2}$$

Problem: A missile of mass $m = 2.3 \times 10^4\, kg$ flies level to the ground at an altitude of d = 10,000m with constant speed $\upsilon = 210 m/s$. What is the magnitude of the missile's angular momentum relative to a point on the ground directly below its flight path?

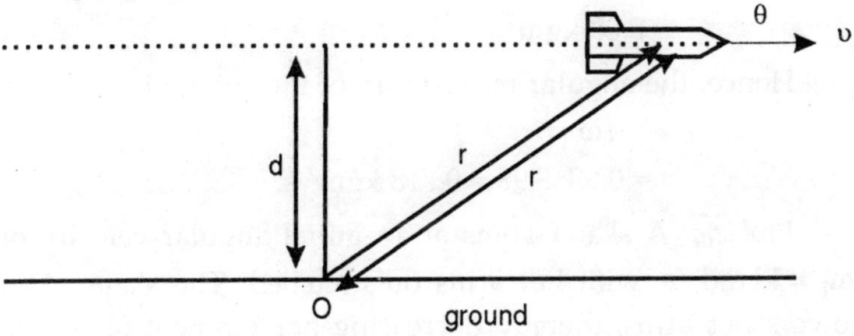

Solution : The missile's angular momentum about point O is

$$L = m\upsilon r \sin\theta$$

where θ is the angle subtended between the missile's velocity vector and its position vector relative to O. However,

$$r\sin\theta = d$$

where is the distance of closest approach of the missile to point O. Hence,

$$L = m\upsilon d$$
$$= (2.3\times10^4)\times210\times(1\times10^4)$$
$$= 4.83\times10^{10}\text{kgm}^2/\text{s}.$$

Problem: A uniform sphere of mass M=5kg and radius a=0.2m spins about an axis passing through its centre with period T=0.7s. What is the angular momentum of the sphere?

Solution : The angular velocity of the sphere is

$$\omega = \frac{2\pi}{T}$$
$$= \frac{2\pi}{0.7}$$
$$= 0.98 \text{ rad./s}.$$

The moment of inertia of the sphere is

$$I = \frac{2}{5}Ma^2$$
$$= 0.4\times5\times(0.2)^2$$
$$= 0.08 \text{ kgm}^2.$$

Hence, the angular momentum of the sphere is

$$L = I\omega$$
$$= 0.08\times8.98 = 0.718 \text{ kgm}^2/\text{s}.$$

Problem: A skater spins at an initial angular velocity of $\omega_1 = 11\,\text{rad./s}$ with her arms outstretched. The skater then lowers her arms, thereby decreasing her moment of inertia by a factor . What is the skater's final angular velocity? Assume that any friction between the skater's skates and the ice is negligible.

Solution: Neglecting any friction between the skates and the ice, we expect the skater to spin with constant angular momentum. The skater's initial angular momentum is

$$L_1 = I_1\omega_1$$

where I_1 is the skater's initial moment of inertia. The skater's final angular momentum is

$$L_2 = I_2\omega_2$$

where $I_2$is the skater's final moment of inertia, and ω_2 is her final angular velocity. Conservation of angular momentum yields $L_2 = L_2$, or

$$\omega_2 = \frac{I_1}{I_2}\omega_2$$

Now, we are told that I_1/I_2=8. Hence, $\omega_2 = 8 \times 11 = 88\, rad./s$

Chapter 6

Elasticity

When externally applied forces are balanced, no net force will act on the object and hence no translational motion can be produced. However, deformation in shape or/and size can occur depending on the intermolecular forces with which the molecules or atoms of the subject hold each other.

In such a case the objects can be classified as follows :

Rigid Objects

If no deformation occurs whatsoever, then the object is called a Rigid Object.

Elastic Objects

Deformation does occur, but it disappears if the balanced external forces are removed.

Plastic Objects

Here, deformation does occur, and it remains permanent even if the balanced external forces are removed.

From the classification the elastic property can be defined as :

If the object is capable of resisting deformation and is capable of overcoming deformation and regaining its original shape and size when externally applied balanced forces are removed then the object possesses the property called elasticity.

TYPES OF DEFORMATIONS

Three types of deformations can be distinguished as follows:

Longitudinal or Tensile Deformation

This is the deformation in length for an object in the form of a wire or a cylinder of negligible cross- sectional area compared to the length when the applied forces are parallel to the length.

Shear Deformation

This is the deformation in area and shape when applied forces are parallel to a surface.

Bulk Deformation

This is the deformation in volume and/or shape when applied forces are normal to each and every point of the surface of the object.

DEFORMING AND RESTORING FORCES

The externally applied balanced forces producing deformation are called deforming forces. The internally balanced forces which arise in response to deforming forces by virtue of elasticity of an object, which resist deformation and restore the object to its original shape and size upon withdrawal of deforming forces, are called restoring forces.

The deformed object remains in equilibrium due to deforming and restoring forces balancing each other.

STRAIN AND STRESS

Strain

The measure of deformation produced is called Strain. For the type of deformation, these can be the following kinds of strains.

Longitudinal or Tensile Strain

The change in length (Δl) to the original length (l) is called tensile strain.

$$\text{i.e., Tensie Strain} = \frac{\Delta l}{l}$$

Shear Strain

The change in area (*DA*) to the original area (*A*) is called shear deformation.

i.e., Shear Strins = $\frac{\Delta A}{A}$

Note : For a small deformation angle of shear (q) is also considered as Shear Strain.

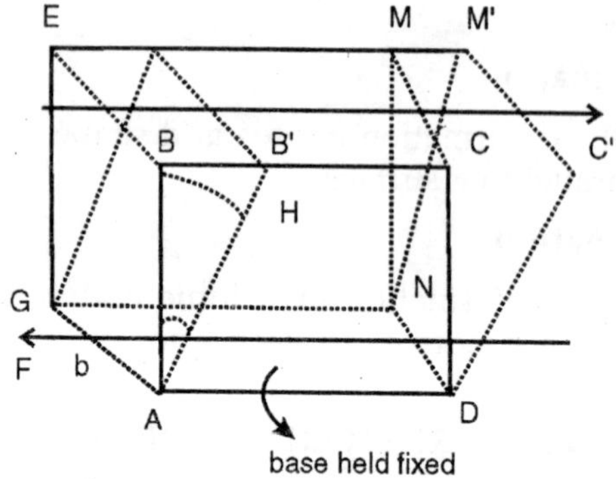

The figure shows Shear deformation of the rectangular block ABCDGEMN, due to displacement of face BCME to position B′C′M′E′ when forces $\vec{F}, -\vec{F}$ *are* applied on top & base as shown. The angle q by which the force ABEG (or DCMN) is twisted to the position AB′E′G (or DC′M′N′) is called angle of shear.

Now,

$$\text{Shear Strain} = \frac{\Delta A}{A}$$

$$= \frac{\text{Area of } AB'E'G - \text{Area of } ABEG}{\text{Area of } ABEG}$$

$$= \frac{(l'-l)\not{b}}{l\not{b}}$$

$$= \frac{AB' - AB}{l}$$

$$= \frac{AB' - AH}{l}$$

$$= \frac{B'H}{l}$$

but, θ is small, $\therefore \; B'H \cong BB'$

$$\therefore \text{ Shear strain} = \frac{BB'}{l} = \frac{BB'}{AB} = \tan\theta$$

$$\therefore \text{ Shear strain} = \tan\theta$$

$\therefore$ Shear strain $\cong \theta$ (because θ is small)

Bulk Strain

The change in the volume (ΔV) to original volume (V) is called bulk strain.

$$\text{i.e. Bulk Strain} = \frac{\Delta V}{V}$$

Note: It is obvious from the definition of Strain, that Strain is just a number without a unit.

Stress

The measure of resistance to deformation is called stress. The following stresses corresponding to each type of deformation can be defined.

Longitudinal or Tensile Stress

The restoring forces per unit normal cross-sectional area is called the tensile stress.

$$\text{i.e. tensile stress} = \frac{\text{Restoring force}}{\text{Area of cross-sections to restoring forces}}$$

$$= \frac{F}{A}$$

Shear Stress

The restoring force, per unit lateral area, is called shear stress.

$$\text{i.e. Shear stress} = \frac{\text{Restoring Force}}{\text{Lateral Area}} = \frac{F}{A}$$

Bulk Stress

The restoring force per unit normal area of the surface is called bulk stress.

$$\text{i.e. Bulk stress} = \frac{\text{Restoring Force}}{\text{Normal Surface Area}} = \frac{F}{A}$$

Since Pressure = Force / Normal Area

$\therefore$ Bulk stress = Pressure

Note : The stress is measured in the unit of Newton/ $(\text{meter})^2$ i.e., N/m^2.

HOOKE'S LAW

The statement regarding strain & stress being proportional to each other in the elastic region is known as Hooke's law. "Within Elastic limit stress and strain are directly proportional to each other".

i.e., Stress a Strain within elastic limit.

$$\text{or } \frac{\text{Stress}}{\text{Strain}} = Constant$$

The constant ratio of stress to strain within elastic limit is called elastic modules.

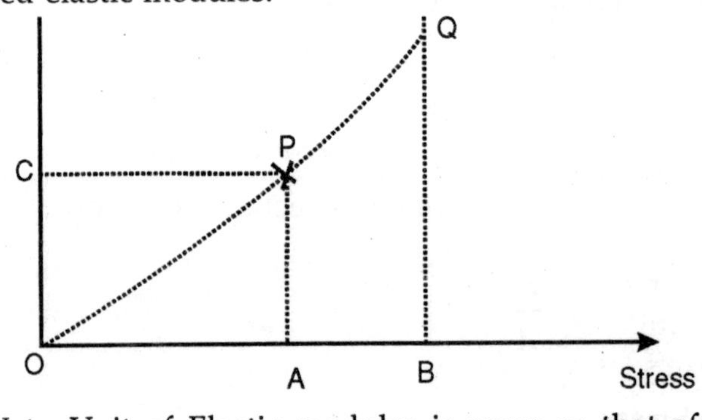

Note: Unit of Elastic modules is same as that of stress (as strain has no limit) viz. n/m^2.

The behaviour of an elastic object, with increasing stress is represented in the Stress - Strain Curve shown below.

OP = Linear Elastic region

PQ = Rising Fatigue region

QR = Total Yield, with breaking into pieces
P = Elastic limit
Q = Yield Point

Linear Elastic Region

In this region if stress is increased from O to A & is decreased back from A to O then, the strain increases in direct proportion from, O to C and decreases from C to O in the same proportion. In this region the object retain its elasticity, strain is the directly proportional to strain. The limit point 'P' of this region is called elastic limit.

Rising Fatigue region

In this strain region increases with the same amount as, the increase in stress; eventually the substance looses its elasticity partially, and therefore becomes partially plastic. This deformation remains permanent, even if the stress is reduced back from B to A and then from A to O. The end point of this region 'Q' is called the Yield Point.

The Total Yield Region

The region QR, represents infinitely large increase in strain produced when stress is increased even by small amount; the substance has completely lost its elasticity & has become completely plastic; hence it yields completely to deforming forces. The substance in this region can be broken into pieces.

ELASTIC MODULES

For the three types of deformation viz. Tensile, Shear and Bulk the elastic modules can also be defined using Hooke's law.

Young's Modulus : Y

The constant ratio of Tensile stress to Tensile Strain, within the Elastic limit, is called the Young's modulus.

$$i.e., Y = \frac{\text{Tensile Stress}}{\text{Tensile Strain}}$$

$$= \frac{FA}{\Delta l / l} n/m^2$$

Modules of Rigidity : η

The constant ratio of shearing stress to shearing strain, within the elastic limit, is called the modulus of Rigidity.

$$\text{i.e., } \eta = \frac{\text{Shearing Stress}}{\text{Shearing Strain}}$$

$$= \frac{F/A}{\Delta a / a}$$

$$= \frac{F/A}{\theta} n/m^2$$

Bulk Modulus : K

The constant ratio of Bulk stress to Bulk Strain, within, the elastic limit, is called the Bulk modulus :

$$i.e., K = \frac{\text{Bulk Stress}}{\text{Bulk Strain}}$$

$$= \left|\frac{\Delta p}{\frac{\Delta v}{v}}\right| n/m^2$$

STRAIN ENERGY

When a wire is stretched, then the work done on the wire is stored in the form of internal P.E. of wire. The amount of stored P.E. per unit volume of wire is called strain energy.

Let a wire of length 'L' be stretched by the means of a force 'F' to the length $(L + x)$; then work done by the force in stretching the wire further by length dx is

$$dW = Fdx$$

$$\text{But, } Y = \frac{F/A}{\Delta l / L}$$

$$= \frac{F/A}{x/L}$$

$$= \frac{FL}{Ax}$$

$$\therefore F = \left(\frac{YA}{L}\right)x$$

$$\therefore dW = \left(\frac{YA}{L}\right)x\,dx$$

$\therefore W$ = Total work done in stretching the wire by length 'x'

$$= \int_0^x \left(\frac{YA}{L}\right)x\,dx$$

$$= \frac{YA}{L}\left(\frac{x^2}{2}\right)_0^x$$

$$= \frac{1}{2}\left(\frac{YA}{L}\right)x\,X\,x$$

$$= \frac{1}{2}F \times x$$

$$\therefore W = \frac{1}{2} \times \text{Force} \times \text{Extension produced by the force}$$

$$\therefore W = \frac{1}{2} \times F \times x$$

$$\therefore \frac{\text{W}}{\text{V}} = \frac{1/2 \times F \times x}{V}$$

$$= \frac{1/2 \times F \times x}{A \times L}$$

$$\therefore \text{ Strain Energy}=\left(\frac{1}{2}\right)\left(\frac{F}{A}\right)\left(\frac{x}{L}\right)$$

$$\therefore \text{ Strain Energy}=\frac{1}{2}(\text{Tensile Stress})(\text{Tensile Strain})$$

$$\text{Now, } Y = \frac{\text{Tensile Stress}}{\text{Tensile Stress}}$$

$$\therefore \text{Tensile Stress} = Y \times \text{Tensile Strain}$$

$$\text{or Tensile Stress} = \frac{\text{Tensile Strain}}{\text{Y}}$$

$$\therefore \text{ Strain Energy} = \frac{1}{2}(Y \times \text{Tensile strain})\text{Tensile strain}$$

$$\therefore \text{ Strain Energy} = \frac{1}{2}Y(\text{Tensile strain})^2$$

$$\text{or Strain Energy} = \frac{1}{2}(\text{Tensile strain}) \times \left(\frac{\text{Tensile stress}}{\text{Y}}\right)$$

$$\therefore \text{ Strain Energy}=\frac{1}{2} \times \frac{(\text{Tensile stress})^2}{\text{Y}}$$

Solved Problems

1. A steel wire of length 2m is hanging from a rigid horizontal support. How much energy is stored in it when a load of 5 Kg is suspended on it ? If the load is increased to 10 Kg then by how much the energy stored will increase ? Y = 2×10^{11} N/m^2, A = Area of cross section = $10^{-6} m^2$

Solution

Given $l = 2m$, $F_1 = 5\times9.8N$, $F_2 = 10\times9.8N$
$A = 10^{-6}m^2$, $Y = 2\times10^{11}\ N/m_2$ W = work done by Force 'F' in extending the length of wire by 'x'.

$$\frac{1}{2}Fx = \frac{1}{2}Fx$$

$$= \frac{1}{2}Fx\times\left(\frac{x}{l}\right)\times l \qquad \text{but } \frac{x}{l} = \text{strain}$$

$$= \frac{1}{2}F\times\text{strain}\times l \qquad \text{but } Y = \frac{\text{stress}}{\text{strain}},$$

$$\therefore \text{strain} = \frac{\text{strees}}{Y}$$

$$= \frac{1}{2}F\times\frac{\text{stress}}{\text{Y}}\times l \qquad \text{but stress} = \frac{F}{A}$$

$$= \frac{1}{2}F\times\frac{1}{\text{Y}}\times\frac{F}{A}\times l$$

$$= \frac{F^2 l}{2AY}$$

$$\therefore W_1 = \frac{F_1^2 l}{2AY}$$

$$= \frac{(5\times9.8)^2\times2}{2\times10^{-6}\times2\times10^{11}}$$

$$W_1 = \frac{(49)^2}{2}\times10^{-5}\,Joule$$

$$\therefore W = 1200.5\times10^{-5}$$

$$\therefore W_1 = 1.2\times10^{-2}\,\text{Joule}$$

Likewise,

$$W_2 = \frac{F_2^2 l}{2AY}$$

$$= \frac{(10\times 9.8)^2 \times 2}{2\times 10^{-6} \times 2\times 10^{11}}$$

$$W_2 = 4.8\times 10^{-2}\ joule$$

∴ Increase in energy stored when $F_1 = 5\times 9.8$ N is increased to $F_2 = 10\times 9.8N$

$$= W_2 - W_1$$

$$= 4.8\times 10^{-2} - 1.2\times 10^{-2}$$

$$= 3.6\times 10^{-2} \text{ Joule}$$

∴ Increase in energy stored = 3.6×10^{-2} Joule

2. An enclosed mass of ideal gas has isothermal bulk modulus equal to its pressure, and has adiabatic bulk modulus equal to Y times pressure where Y = ratio of molar Cp to Cv of the gas. State true or false and justify your answer.

Solution

Given K = Bulk modulus = $\left|V\frac{dp}{dV}\right|$

1. Isothermal Bulk modulus K_1 for an ideal gas

$PV=C$ (under Isothermal change)

$$P\frac{dp}{dV} + V\frac{dp}{dV} = 0$$

$$V\frac{dp}{dV} = -P$$

$$\left|V\frac{dp}{dV}\right| = P$$

$$\therefore KI = P$$

2 Adiabatic Bulk modulus K_A for an ideal gas

$$PV^{\gamma} = C$$

$$\therefore P_{\gamma}V^{\gamma-1} + V^{\gamma}\frac{dp}{dV} = 0$$

$$V\frac{dp}{dV} = -\gamma P$$

$$\left|V\frac{dp}{dV}\right| = \gamma P$$

$$K_A = \gamma p$$

Chapter 7

Kinematics

When an object changes its position relative to a chosen reference object (Frame of reference) it is in motion relative to that 'frame'. Description of motion requires knowledge of the following terms :

DEFINITIONS

1. Displacement($\vec{s}$) : Distance traveled in a given direction is called displacement. Its units 1m.
2. Speed(V) : Distance traveled per unit time or the rate at which distance is covered is called speed Its unit is 1m/s.
3. Velocity($\vec{v}$): Displacement per unit time or rate at which displacement occurs is called velocity. Its unit is 1m/s.
4. Acceleration($\vec{a}$): The rate of change in velocity is called the acceleration. Its unit is $1m/s^2$.

   The object moves from the point P at $t = 0$ sec to the point Q in interval $\delta\tau$ sec.

Consider,

$$\vec{r_i} = \overrightarrow{OP} = \text{Initial Position Vector}$$

$$\vec{r_f} = \overrightarrow{OQ} = \text{Final Position Vector}$$

$$\overrightarrow{\Delta s} = \overrightarrow{PQ} = \vec{r_f} - \vec{r_i} = \text{Displacement Vector}$$

$\overrightarrow{v_i}$: Initial velocity vector

$\overrightarrow{v_f}$: Initial velocity vector

$\vec{a}$: Acceleration vector

then $\overrightarrow{v_i} = \dfrac{\Delta\vec{s}}{\Delta t}$ (by definition)

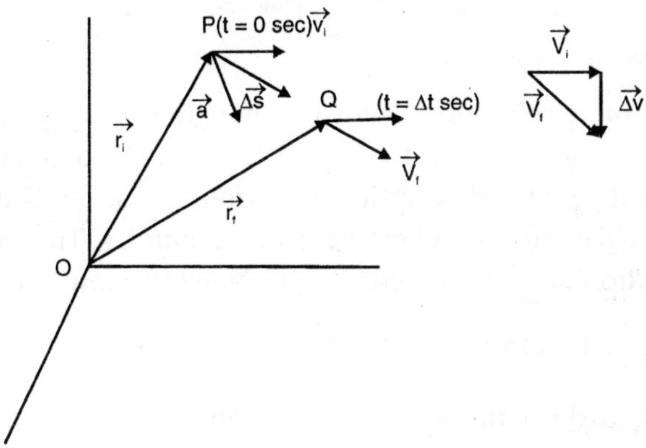

If the time interval Δt is sufficiently large then the velocity obtained is called average velocity. For as accurate description of motion it is necessary to choose Δt as small as possible i.e. $\Delta t \to 0$; then the velocity obtained is called instantaneous velocity

Thus $\vec{v}_{average} = \dfrac{\Delta\vec{s}}{\Delta t}$ Δt large

& $\vec{v}$ = Instantaneous velocity = $\lim\limits_{\Delta t \to 0} \dfrac{\overrightarrow{\Delta s}}{\Delta t} = \dfrac{\overrightarrow{ds}}{dt}$

If $\vec{v}$ = constant, then it is called uniform velocity.

Formotion with uniform velocity,

$\vec{v}_{\text{average}} = \vec{v}_{\text{instantaneous}} = \vec{v}_{\text{uniform}}$

As the displacements must be directly proportional to the interval of time hence the ratio of displacement to time interval $\dfrac{(\overrightarrow{\Delta s})}{(\Delta t)}$ will be constant whether Δt is sufficiently small or arbitrarily large.

Similar considerations also apply to acceleration also

i.e. $\vec{a}_{average} = \dfrac{\overrightarrow{\Delta v}}{\Delta t}$ (Δt large)

$\vec{a}$ instantaneous = $\lim\limits_{\Delta t \to 0} \dfrac{\overrightarrow{\Delta v}}{\Delta t}$

and $\vec{a}_{\text{average}} = \vec{a}_{\text{instantaneous}} = \vec{a}_{\text{uniform}}$,

when $\vec{a}$ = constant

Even if acceleration keeps on changing with time, we do not find it necessary to consider the rate of change in acceleration for the description of motion, and so on. The reason for this will be found when we study dynamics. The causes of acceleration are forces according to Newton's laws of motion.

MOTION IN ONE DIMENSION

Motion under Uniform Acceleration

Let a particle move along X-axis (chosen for convenience as the motion is one dimensional) with uniform acceleration a

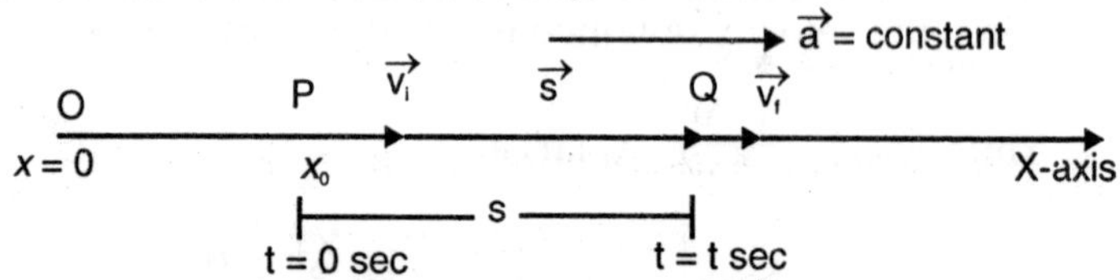

For uniformly accelerated motion, velocity and position can be determined at any instant. Thus motion is accurately and completely described.

Now $\vec{a} = \dfrac{\overrightarrow{dv}}{\Delta t}$ ($\because$ uniform or constant acceleration)

$\therefore\ \Delta\vec{v} = \vec{a} \times \Delta t$

$\therefore\ \overrightarrow{v_f} - \overrightarrow{v_i} = \vec{a} \times (t - 0)$

$\therefore\ \overrightarrow{v_f} = \overrightarrow{v_i} + \overrightarrow{at}$

This motion occurs with changing velocity as is clear from equation, however, it can be regarded as motion with uniform velocity if the average of $\overrightarrow{v_i}$ & $\overrightarrow{v_f}$ it is taken as its uniform velocity

$$\text{therefore } \vec{v}_{average} = \frac{\vec{s}}{t}$$

$$\text{but } \vec{v}_{average} = \frac{\overrightarrow{v_i} + \overrightarrow{v_f}}{2}$$

$$\therefore \frac{\overrightarrow{v_i} + \overrightarrow{v_f}}{2} = \frac{\vec{s}}{t}$$

$$\therefore \vec{s} = \left(\frac{\overrightarrow{v_i} + \overrightarrow{v_f}}{2}\right)t$$

$$\therefore \vec{s} = \left(\frac{\overrightarrow{v_i} + \overrightarrow{v_i} + \overrightarrow{at}}{2}\right)t$$

$$= \left(\overrightarrow{v_i} + \frac{1}{2}\overrightarrow{at}\right)t$$

$$\therefore \vec{s} = \overrightarrow{v_i}.t + \frac{1}{2}\overrightarrow{at}^2$$

$$\text{But } \vec{s} = \vec{x} - \vec{x}_0$$

$$\vec{x} = \vec{x}_0 + \vec{v}_i.t + \frac{1}{2}\overrightarrow{at^2}$$

Consider

$$\overrightarrow{v_f} \cdot \overrightarrow{v_f} = \left(\overrightarrow{v_i} + \overrightarrow{at}\right)\cdot\left(\overrightarrow{v_i} + \overrightarrow{at}\right)$$

$$\therefore \overrightarrow{v_f} \cdot \overrightarrow{v_f} = \overrightarrow{v_i}.\overrightarrow{v_i} + 2\overrightarrow{v_i}\overrightarrow{at} + \left(\overrightarrow{a\,t}\right)\cdot\left(\overrightarrow{at}\right)$$

$$\therefore vf^2 = \overrightarrow{v_i^2} + 2a\cdot\left[\overrightarrow{v_i}t + \frac{1}{2}\overrightarrow{at^2}\right]$$

$$\therefore \; vf^2 = v_i^2 + 2\vec{a}\cdot\vec{s}$$

$$\therefore vf^2 = vi^2 + 2\vec{a}.\vec{s}$$

The equations are called linear kinematical equations of motion under uniform acceleration.

The vector form of the equation has an advantage as the conventional +ve & –ve sign for vectors pointing to the right & left respectively need not be kept in mind. The equations may also be represented in the scalar form also as

$$\left.\begin{aligned} v_f &= v_i + at \\ s &= v_i t + 1/2ai^2 \\ vf^2 &= v_i^2 + 2as \end{aligned}\right]$$ but the proper signs for the quantities u, v, a, s have to beused

Graphical Representations of Displacement, Velocity, Acceleration

i. One dimensional uniformly accelerated motion.

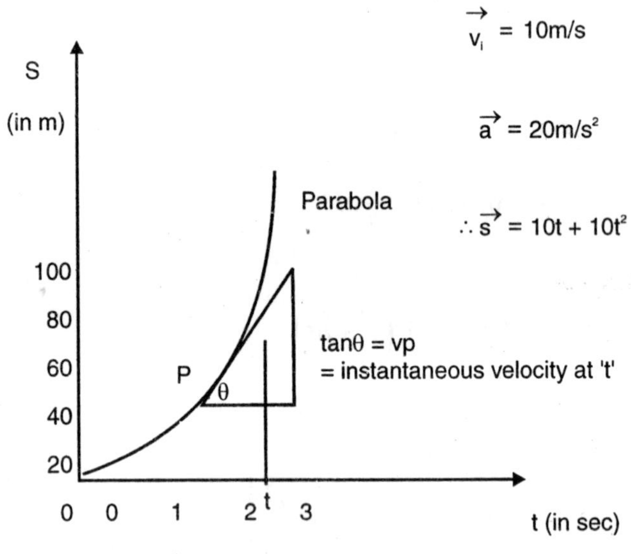

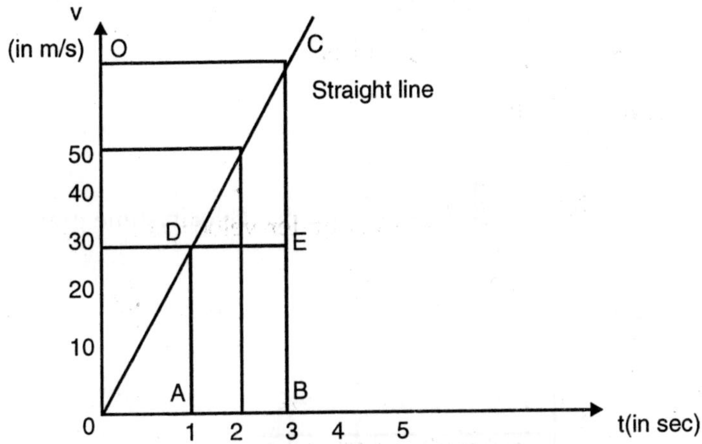

Area ABCD = Area of Rectangle ABED + Area of DEC

$$= AB \times AD + \frac{1}{2} DE \times EC$$

$$= (3-1) \times (30-0) + \frac{1}{2} (3-1) \times (70-30)$$

$$= 60 + 40$$

$$= 100 \text{ m.}$$

Now, displacement between 1st to 3rd sec is

$$= 10 \times 3 + \frac{1}{2} \times 10 \times 32 - \{10 \times 1 + \frac{1}{2} \times 20 \times 1^2\}$$

$$= 10 [3 + 9] - \{ 10 (1 + 1) \}$$

$$= 10 \{12 - 2\}$$

$$= 10 \{10\}$$

$$= 100 \text{ m}$$

Thus displacement in a given interval is

= Area under velocity-time graph for the given interval.

This result is obvious, as can be seen from

$$v = \frac{ds}{dt} \text{ (in scalar form)}$$

$$\therefore ds = v\,dt$$

$$\therefore \int_{s_1}^{s_2} ds = \int_{t_1}^{t_2} v dt = \text{Area under velocity-time graph}$$

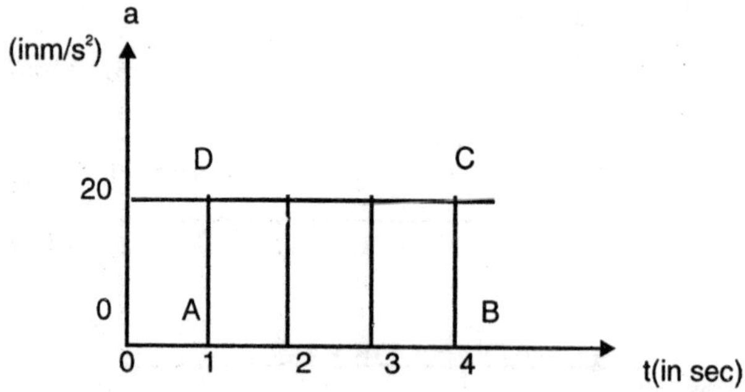

$$\int_1^4 a.dt = \int_1^4 \frac{dv}{dt} dt = \int_1^4 dv = v_4 - v_1$$

$$\therefore a \int_1^4 dt = v_4 - v_1$$

$$\therefore a(4-1) = v_4 - v_1$$

$$\therefore v_4 - v_1 = 3 \times 30 = 60\text{m/s}$$

$$\therefore \int_1^4 adt = \int_1^4 dv$$

$$= \text{Area under the acceleration-time graph.}$$

The above results can be generalized when the acceleration is not uniform.

Thus, to summarize

i. v = slope of tangent to s-t graph

ii. Δs = Area under v-t graph

iii. a = slope of tangent to v-t graph.

iv. Δv = Area under a-t graph.

MOTION IN TWO DIMENSIONS

Projectile Motion

The motion of an object projected at an angle to the horizontal level and subjected to acceleration due to the gravity of the Earth alone, is called projectile motion. The air resistance is assumed to be absent and the horizontal level is considered to be a plane and the curvature of Earth's surface is neglected.

Locally, the projectile motion can be analyzed as a two component motion : Horizontal component with uniform horizontal velocity and vertical component with uniform acceleration due to gravity of Earth.

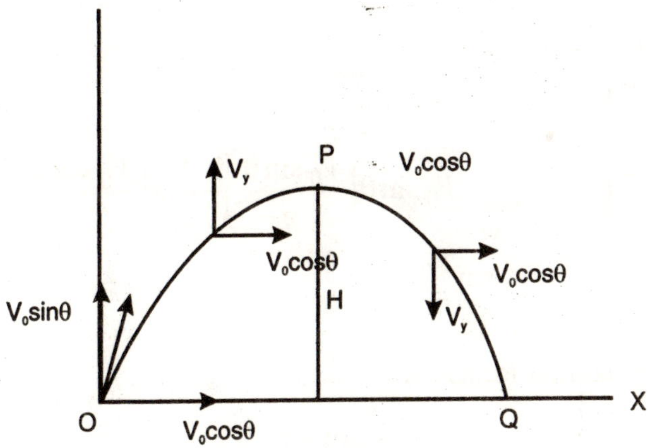

Hor. Component : $V_x = v_0 \cos\theta$

$$x = v_0 \cos\theta t$$

Ver. Component: $V_y = V_0\sin\theta - gt$

$$s = Vit + \frac{1}{2}gt^2$$

$$y = v_0 \sin\theta t - \frac{1}{2} g t^2$$

Time to rise up to highest point P:(T)

At P, $V_y = 0$

$$0 = V_0 \sin\theta - gt$$

$$T = \frac{V_0 \sin\theta}{g}$$

Time of flight: τ

τ = Time to rise up to P from O
+ Time to fall down from P to Q
= T + T (Rising and falling time is same under 'g')
= 2T

$$= \frac{2V_0 \sin\theta}{g}$$

Maximum height: H

H = y_{max}, occurs at t = T

$$\therefore H = y_{max} = V_0 \sin\theta\left(\frac{V_0 \sin\theta}{g}\right) - \frac{1}{2} g\left(\frac{V_0 \sin\theta}{g}\right)^2$$

$$\therefore H = \frac{v_0^2 \sin^2\theta}{2g}$$

Horizontal Range: R

R = x_{max}, occurs at t = τ

$$R = V_0 \cos\theta \times \tau$$

$$\therefore R = V_0 \cos\theta \times \frac{2V_0 \sin\theta}{g}$$

$$= \frac{v_0^2 \sin 2\theta}{g}$$

$$\therefore R = \frac{v_0^2 \sin 2\theta}{g}$$

Now $\sin(180° - 2\theta) = \sin 2\theta$

$$\therefore R = \frac{v_0^2 \sin(180° - 2\theta)}{g}$$

Let, $2\alpha = 180° - 2\theta$

then $\alpha = 90° - \theta$

With given velocity V_0 there are two angles of projection θ & $90° - \theta$ for which the horizontal range remains same; the vertical heights however will be different. Also if $\theta = 45°$ then $\sin 2\theta = \sin 90° = 1$ (maximum)

$R = \frac{v_0^2}{g}$ is the maximum range

possible at $\theta = 45°$; then $a = 90° - \theta = 90° - 45° = 45°$

These results are represented in figure.

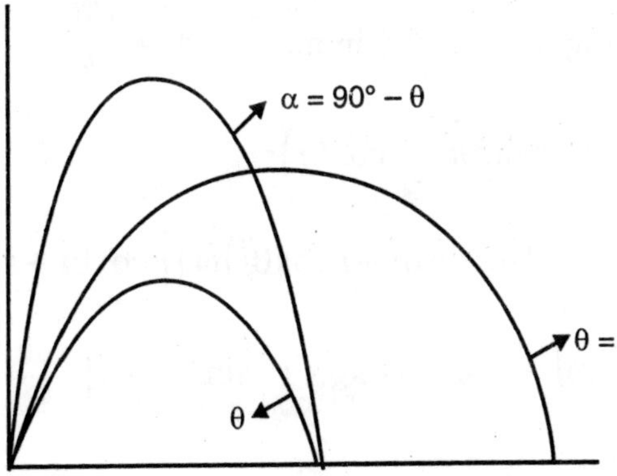

Uniform Circular Motion :

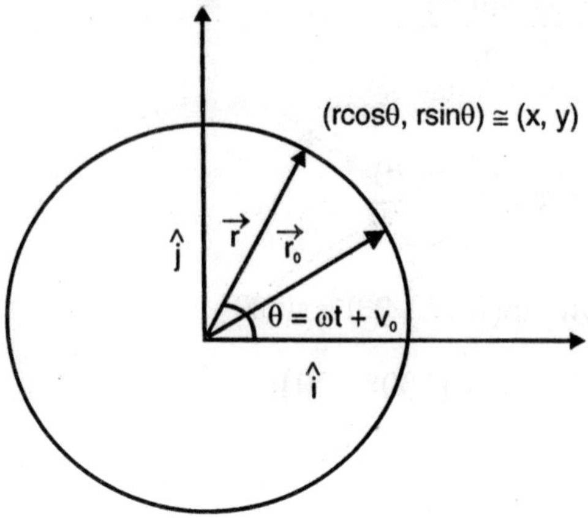

$\vec{r} = x\hat{i} + y\hat{j}$

The circular motion with uniform linear speed is called U.C.M.

$$\vec{r} = r\cos\theta\hat{i} + r\sin\theta\hat{j}$$

$$\therefore \vec{v} = \frac{\overrightarrow{dr}}{dt} = -r\sin\theta\frac{d\theta}{dt}\hat{i} + r\cos\theta\frac{d\theta}{dt}\hat{j}$$

$$\vec{v} = \left(-r\sin\theta\hat{i} + r\cos\theta\hat{j}\right)\frac{d\theta}{dt}$$

put ω = angular velocity in magnitude = $\frac{d\theta}{dt}$

$$\vec{v} = \left(-r\sin\theta\hat{i} + r\cos\theta\hat{j}\right)\omega$$

consider $\vec{v}\cdot\vec{r} = \left(-r\sin\theta\hat{i} + r\cos\theta\hat{j}\right)\omega.\left(r\cos\theta\hat{i} + r\sin\theta\hat{j}\right)$

$$= \omega\left(-r^2\sin\theta\cos\theta + r^2\sin\theta\cos\theta\right)$$

$$\therefore v \perp^r \vec{r} = 0$$

Linear velocity v is along the tangent to the circle.

Consider $|\vec{v}| = v = \sqrt{(-r\omega\sin\theta)^2 + (r\omega\cos\theta)^2}$

$$\therefore \text{v} = \sqrt{(r\omega)^2}$$

$$\therefore \text{v} = r\,\omega$$

$$\therefore \omega = \frac{v}{r}$$

But v = constant in u.c.m, therefore angular velocity magnitude ω is aslo constant. Since direction of $\vec{\omega}$ is parallel to $\hat{k}$ for u.c.m of a particle in XY- plane therefore $\vec{\omega}$ is itself constant in u.c.m.

$$\text{Now, } \omega = \frac{d\theta}{dt}$$

$$\omega = \omega t + \theta_0$$

Eqn can can be re-written as

$$\vec{r} = r\cos(\omega t + \theta_0)\hat{i} + r\sin(\omega t + \theta_0)\hat{j}$$

Consider

$$\vec{a} = \frac{\overrightarrow{dv}}{dt}$$

$$= \frac{d}{dt}\left[-\omega r\sin\theta\hat{i} + r\cos\theta\hat{j}\right]$$

$$= -\omega r\cos\theta\frac{d\theta}{dt}\times\hat{i} - \omega r\sin\theta\frac{d\theta}{dr}\times\hat{j}$$

$$= -\omega^2\left(r\cos\theta\hat{i} + r\sin\theta\hat{j}\right)$$

$$= -\omega^2\vec{r}$$

$$\vec{a} = -\omega^2\vec{r}$$

Thus linear acceleration $\vec{a}$ is directed along the radius and towards the centre, this is known as the centripetal acceleration.

The magnitude of a is

$$a = \omega^2 \vec{r}$$

$$= \frac{v^2}{r} r$$

$$= \frac{v^2}{r}$$

$$a = \frac{v^2}{r} = \omega^2 r$$

The u.c.m can therefore be depicted.

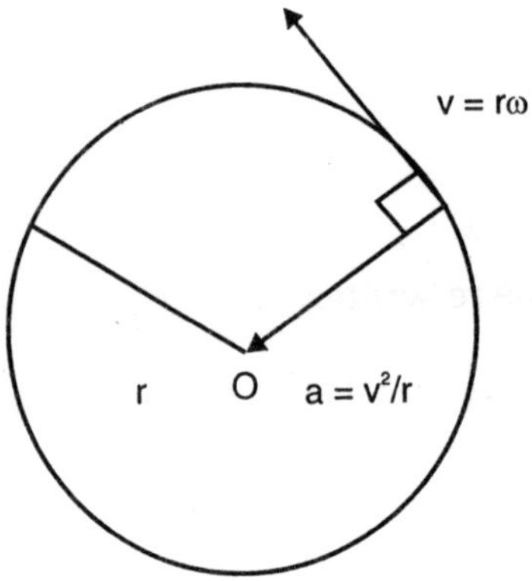

In u.c.m.

$$\omega = \frac{2\pi}{T} = 2\pi n$$

where T = period and n = frequency

Thus,

$$v = \omega r = r \times \frac{2\pi}{T} = r \times 2\pi n$$

$$a = r\omega^2 = \frac{v^2}{r}$$

Solved Problems

Problem: A monkey is sitting on a tree along the line of sight of a riffle which is aimed at the monkey. The moment the rifle is fired, the monkey just jumps off in a vertical direction. Will the bullet hit the monkey ?

Solution

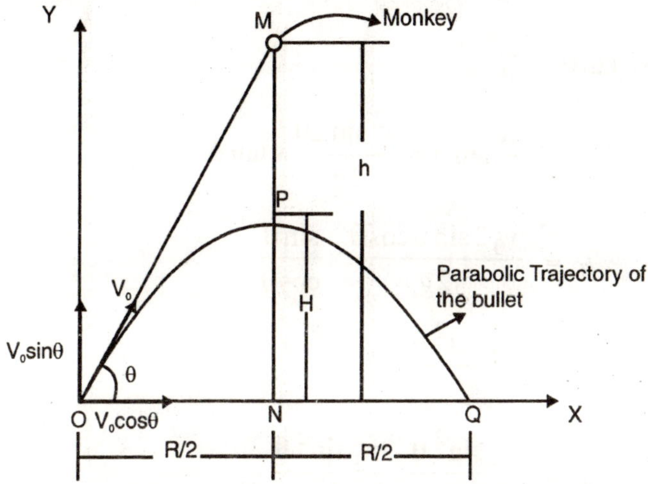

$$H = \text{Maximum Height of Projectile} = \frac{v_0^2 \sin^2 \theta}{2g}$$

$$R = \text{Horizontal Range of Projectile} = \frac{v_0^2 \sin^2 2\theta}{g}$$

$$\tau = \text{Time to rise up to P} = \frac{v_0 \sin \theta}{g}$$

The monkey jumping off at t = 0 will descend down by distance,

$$s = \frac{1}{2} g \tau^2 \qquad \text{if } t = \tau$$

$$\therefore s = \frac{1}{2} g \left(\frac{v_0 \sin \theta}{g} \right)^2$$

$$= \frac{v_0^2 \sin^2\theta}{2g}$$

$\therefore$ After t = τ the monkey is at the height $h - s = h - \frac{v_0^2 \sin^2\theta}{2g}$ of from the ground.

$$\text{Now, } \tan\theta = \frac{h}{R/2}$$

$$\therefore h = \frac{R}{2}\tan\theta = \frac{v_0^2 \sin 2\theta}{2g} \times \tan\theta$$

$$\therefore h = \frac{v_0^2 2\sin\theta\cos\theta}{2g} \times \frac{\sin\theta}{\cos\theta}$$

$$h = \frac{v_0^2 \sin^2\theta}{g}$$

$$\therefore h - s = \frac{v_0^2 \sin^2\theta}{g} - \frac{v_0^2 \sin^2\theta}{2g}$$

$$\therefore h - s = \frac{v_0^2 \sin^2\theta}{2g}$$

$$\text{but } H = \frac{v_0^2 \sin^2\theta}{2g}$$

$$\therefore h - s = H$$

Thus in the interval t the bullet rises to position P and the monkey falls to position P.

Promlem: A simple pendulum of length l m is held in the horizontal position initially, and then it is released. Find the tension in the string at any position and show that the bob has three times its normal weight when it is in the lowest position.

Solution

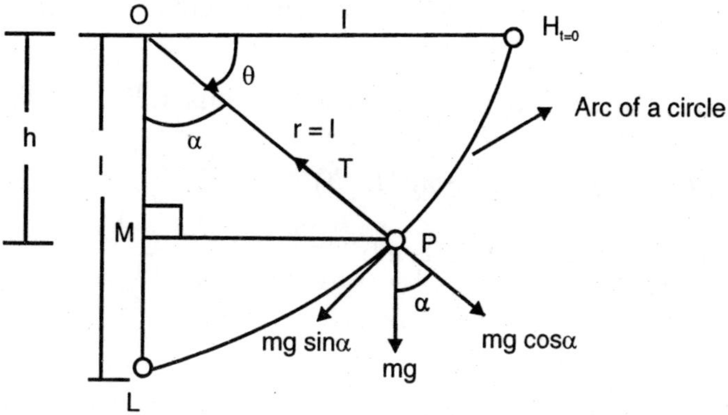

$l = 1$ m.

T = Tension in the string

$\alpha + \theta = 90°$

$t = 0, \theta = 0°$

L : Lowest position

H : Highest position

P : Intermediate position

The bob describes vertical circular motion :

$$\therefore \text{Centripetal force at P} = \frac{mv_p^2}{r} = \frac{mv_p^2}{l}$$

But net force along the string is T – mgcos α, which must act as the centripetal force.

$$\therefore T - mg\cos\alpha = \frac{mv_p^2}{l}$$

$$\therefore T = \frac{mv_p^2}{l} + mg\cos\alpha$$

$$= \frac{mv_p^2}{l} + mg\sin\theta$$

Now, as the bob moves on the arc of a circle from H to P its speed increases from zero to vp and its Potential energy (P.E.) decreases from mg$l$ to mg(l – h)

$\therefore$ gain in Kinetic energy (K.E.) = loss in P.E.

$$\therefore \frac{1}{2}mv_p^2 - 0 = mgl - mg(l-h)$$

(taking P.E.= 0 at 'L' level)

$$\therefore \frac{1}{2}mv_p^2 = mgh$$

$$\therefore \frac{mv_p^2}{l} = \frac{2mgh}{l}$$

$$\text{But } \frac{h}{l} = \frac{OM}{OP} = \cos\alpha$$

$$\therefore \frac{h}{l} = \cos\alpha = \sin\theta$$

$$\therefore \frac{mv_p^2}{l} = 2mg\sin\theta$$

$$T = \frac{mv_p^2}{l} + mg\sin\theta$$

$$= 2mg\sin g\theta + mg\sin\theta$$

$$= 3mg\sin\theta$$

$$\therefore T = 3mg\sin\theta$$

$$\therefore T_L = \text{Tension lowest position L}$$

$$= 3\text{ mg } \sin 90° \quad (\theta = 90° = \angle HOL)$$

$$= 3\text{mg}$$

$$\therefore T_L = 3\text{mg}$$

MOTION IN 3 DIMENSIONS

The purpose of this section is to generalize the previously introduced concepts of *displacement*, *velocity*, and *acceleration* in order to deal with motion in 3 dimensions.

Cartesian Coordinates

Our first task, when dealing with 3-dimensional motion, is to set up a suitable coordinate system. The most straightforward type of coordinate system is called a *Cartesian system*, after René Descartes. A Cartesian coordinate system consists of three mutually perpendicular axes, the x–, y–, and z–axes.

By convention, the orientation of these axes is such that when the index finger, the middle finger, and the thumb of the right-hand are configured so as to be mutually perpendicular, the index finger, the middle finger, and the thumb can be aligned along the $-x$, $-y$, and z–axes, respectively. Such a coordinate system is termed *right-handed*. The point of intersection of the three coordinate axes is termed the *origin* of the coordinate system.

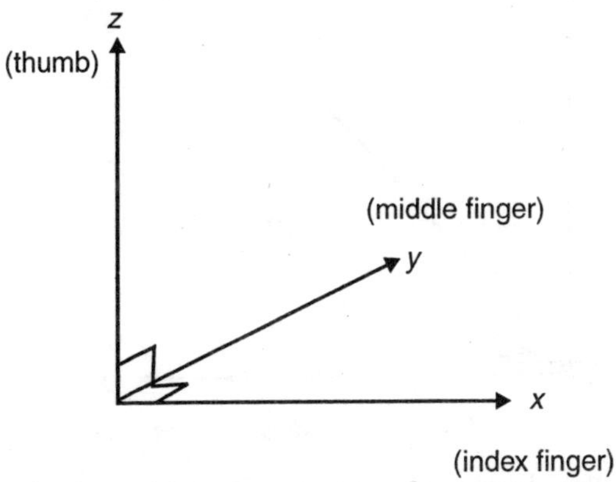

Fig. A right-handed Cartesian coordinate system

VECTOR DISPLACEMENT

Consider the motion of a body moving in 3 dimensions. The body's instantaneous position is most conveniently specified by giving its displacement from the origin of our coordinate system. Note, however, that in 3 dimensions such a displacement possesses both *magnitude* and *direction*. In other words, we not only have to specify how far the body is situated from the origin, we also have to specify in which direction it lies. A quantity which possesses both magnitude and direction is termed a *vector*. By contrast, a quantity which possesses only magnitude is termed a *scalar*. Mass and time are scalar quantities. However, in general, displacement is a vector.

The vector displacement *r* of some point *R* from the origin *O* can be visualized as an arrow running from point *O* to point *R*. Note that in typeset documents vector quantities are conventionally written in a *bold-faced* font (*e.g.*, *r*) to distinguish them from scalar quantities. In free-hand notation, vectors are usually *under-lined*.

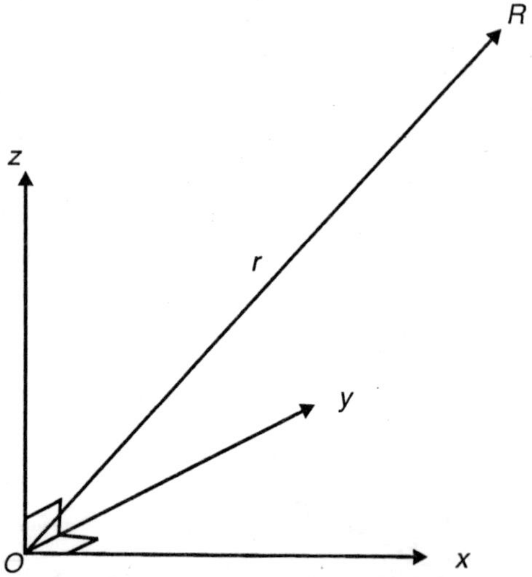

Fig. *A vector displacement*

The vector displacement *r* can also be specified in terms of its *coordinates*:

$$r = (x, y, z)$$

The expression is interpreted as follows: in order to get from point *O* to point *R*, first move *x* meters along the *x*-axis (perpendicular to both the *y*- and *z*-axes), then move *y* meters along the y-axis (perpendicular to both the x- and z-axes), finally move z meters along the z-axis (perpendicular to both the x- and y-axes). Note that a *positive* value is interpreted as an instruction to move x meters along the x-axis in the direction of *increasing x*, whereas a *negative* x value is interpreted as an instruction to move $|x|$ meters along the x-axis in the opposite direction, and so on.

Vector Addition

Suppose that the vector displacement r of some point R from the origin O is specified as follows:

$$r = r_1 + r_2$$

In order to get from point O to point R, we first move from point O to point S along vector r_1, and we then move from point S to point R along vector r_2. The net result is the same as if we had moved from point O directly to point R along vector r. Vector r is termed the *resultant* of adding vectors r_1 and r_2.

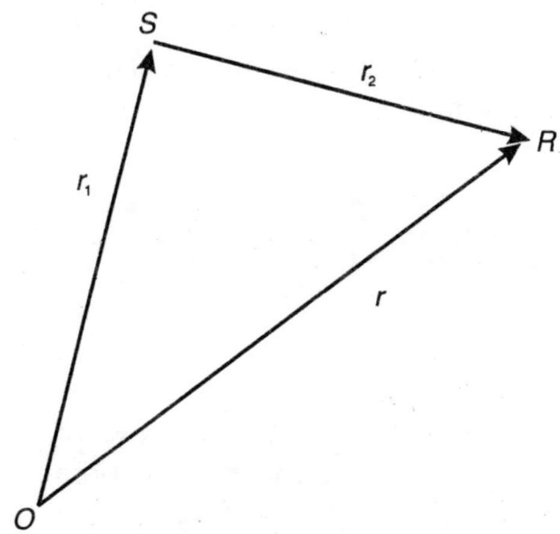

Fig. Vector addition

Note that we have two ways of specifying the vector displacement of point S from the origin: we can either write r_1 or $r - r_2$. The expression $r - r_2$ is interpreted as follows: starting at the origin, move along vector r in the direction of the arrow, then move along vector r_2 in the *opposite* direction to the arrow.

In other words, a *minus* sign in front of a vector indicates that we should move along that vector in the *opposite* direction to its arrow.

Suppose that the components of vectors r_1 and r_2 are (x_1, y_1, z_1) and, (x_2, y_2, z_2) respectively. As is easily demonstrated, the components (x, y, z) of the resultant vector $r = r_1 + r_2$ are

$$x = x_1 + x_2$$

$$y = y_1 + y_2$$

$$z = z_1 + z_2$$

VECTOR MAGNITUDE

If $r = (x, y, z)$ represents the vector displacement of point *R* from the origin, what is the distance between these two points? In other words, what is the length, or *magnitude*, $r = |r|$, of vector *r*. It follows from a 3-dimensional generalization of Pythagoras' theorem that

$$r = \sqrt{x^2 + y^2 + z^2}$$

Note that if $r = r_1 + r_2$ then

$$|r| \leq |r_1| + |r_2|.$$

In other words, the magnitudes of vectors cannot, in general, be added algebraically. The only exception to this rule occurs when the vectors in question all point in the same direction. According to inequality, if we move 1m to the North and next move 1m to the West then, although we have moved a total distance of 2m, our net distance from the starting point is less than 2m—of course, this is just common sense.

Scalar Multiplication

Suppose that $s = \lambda r$. This expression is interpreted as follows: vector *S* points in the *same* direction as vector *r*, but the length of the former vector is $|\lambda|$ times that of the latter. Note that if is negative then vector points in the *opposite* direction to vector, and the length of the former vector is times that of the latter. In terms of components:

$$s = \lambda(x, y, z) = (\lambda x, \lambda y, \lambda z)$$

In other words, when we multiply a vector by a scalar then the components of the resultant vector are obtained by multiplying *all* the components of the original vector by the scalar.

Diagonals of a Parallelogram

The use of vectors is very well illustrated by the following rather famous proof that the diagonals of a parallelogram mutually bisect one another.

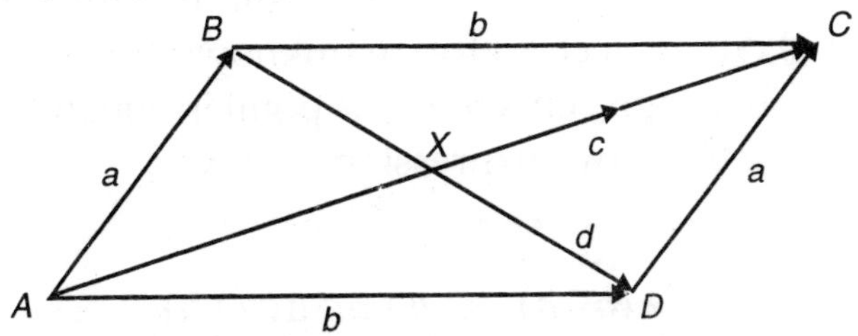

Fig. A parallelogram

Suppose that the quadrilateral ABCD in figure is a parallelogram. It follows that the opposite sides of ABCD can be represented by the *same* vectors, *a* and *b*: this merely indicates that these sides are of equal length and are parallel (*i.e.*, they point in the same direction). Note that figure illustrates an important point regarding vectors. Although vectors possess both a magnitude (length) and a direction, they possess no intrinsic position information. Thus, since sides *AB* and are *DC* parallel and of equal length, they can be represented by the *same* vector *a*, despite the fact that they are in different places on the diagram.

The diagonal *BD* in figure can be represented vectorially as $d = b - a$. Likewise, the diagonal *AC* can be written $c = a + b$.

The displacement x of the centroid X from point A can be written in one of two different ways:

$$x = a+\lambda d$$

$$x = b+a-\mu c$$

equation is interpreted as follows: in order to get from point A to point X, first move to point B(along vector a), then move along diagonal BD(along vector d) for an unknown λ fraction of its length. Equation is interpreted as follows: in order to get from point A to point X, first move to point B(along vector a), then move along diagonal BD(along vector d) for an unknown fraction λ of its length. Equation is interpreted as follows: in order to get from point to point, first move to point (along vector), then move to point (along vector), finally move along diagonal (along vector) for an unknown fraction of its length. Since represents the *same* point in equations, we can equate these two expressions to give

$$a+\lambda(b-a) = b+a-\mu(a+b)$$

Now vectors a and b point in *different* directions, so the only way in which the expression can be satisfied, in general, is if the coefficients of a and b match on either side of the equality sign. Thus, equating coefficients of a and b, we obtain

$$1-\lambda = 1-\mu$$

$$\lambda = 1-\mu$$

It follows that $\lambda=\mu=1/2$. In other words, the X centroid is located at the halfway points of diagonals BD and AC: *i.e.*, the diagonals mutually bisect one another.

VECTOR VELOCITY AND VECTOR ACCELERATION

Consider a body moving in 3 dimensions. Suppose that we know the Cartesian coordinates, x, y, and z, of this body as time, *t*, progresses. Let us consider how we can use this information to determine the body's instantaneous velocity and acceleration as functions of time.

The vector displacement of the body is given by

$$r(t) = [x(t), y(t), z(t)]$$

By analogy with the 1-dimensional equation, the body's vector velocity $v = (v_x, v_y, v_z)$ is simply the *derivative* of with respect to *t*. In other words,

$$v(t) = \lim_{\Delta t \to 0} \frac{r(t+\Delta t) - r(t)}{\Delta t}$$

When written in component form

$$v_x = \frac{dx}{dt}$$

$$v_y = \frac{dy}{dt}$$

$$v_z = \frac{dz}{dt}$$

Thus, the *x*-component of velocity is simply the time derivative of the *x*-coordinate, and so on.

By analogy with the 1-dimensional equation, the body's vector acceleration $a = (a_x, a_y, a_z)$ is simply the *derivative* of *v* with respect to *t*. In other words,

$$a(t) = \lim_{\Delta t \to 0} \frac{v(t+\Delta t) - v(t)}{\Delta t} = \frac{dv}{dt} = \frac{d^2 r}{dt^2}$$

When written in component form

$$a_x = \frac{dv_x}{dt} = \frac{d^2 x}{dt^2}$$

$$a_y = \frac{dv_y}{dt} = \frac{d^2 y}{dt^2}$$

$$a_z = \frac{dv_z}{dt} = \frac{d^2 z}{dt^2}$$

Thus, the x-component of acceleration is simply the time derivative of the x-component of velocity, and so on.

As an example, suppose that the coordinates of the body are given by

$$x = \sin t$$

$$y = \cos t$$

$$z = 3t$$

The corresponding components of the body's velocity are then simply

$$v_x = \frac{dx}{dt} = \cos t$$

$$v_y = \frac{dy}{dt} = -\sin t$$

$$v_z = \frac{dz}{dt} = 3$$

whilst the components of the body's acceleration are given by

$$a_x = \frac{dv_x}{dt} = -\sin t$$

$$a_y = \frac{dv_y}{dt} = -\cos t$$

$$a_z = \frac{dv_z}{dt} = 0$$

Motion with Constant Velocity

An object moving in 3 dimensions with constant velocity v possesses a vector displacement of the form

$$r(t) = r_0 + vt$$

where the constant vector r_0 is the displacement at time t = 0. Note that $dx/dt = v$ and $d^2r/dt^2 = 0$, as expected. The object's trajectory is a *straight-line* which passes through point r_0 at time $t = 0$ and runs parallel to vector v.

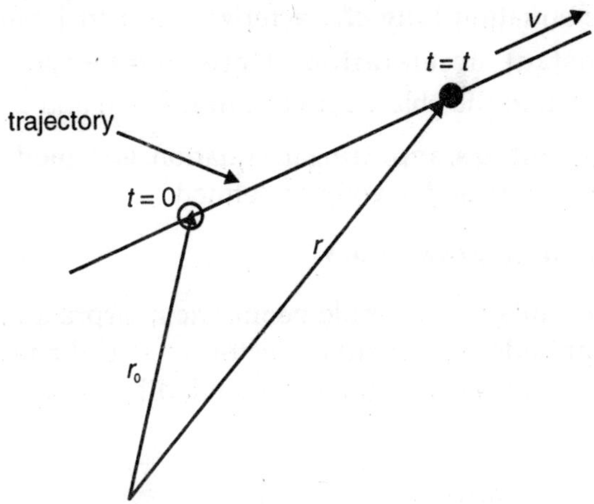

Fig. Motion with constant velocity

Motion with Constant Acceleration

An object moving in 3 dimensions with constant acceleration a possesses a vector displacement of the form

$$r(t) = r_0 + v_0 t + \frac{1}{2} a t^2$$

Hence, the object's velocity is given by

$$v(t) = \frac{dr}{dt} = v_0 + at$$

Note that $dv/dt = a$, as expected. The constant vectors r_0 and v_0 are the object's displacement and velocity at time t = 0, respectively.

As is easily demonstrated, the vector equivalents of Equations are:

$$s = v_0 t + \frac{1}{2} a t^2$$

$$v = v_0 + at$$

$$v^2 = v_0^2 + 2a.s.$$

These equation fully characterize 3-dimensional motion with constant acceleration. Here, $s = r - r_0$ is the net displacement of the object between times $t = 0$ and t.

The quantity $a.s$, appearing in equation, is termed the *scalar product* of vectors and, a and s is defined

$$a.s = a_x s_x + a_y s_y + a_z s_z$$

The formula has a simple geometric interpretation. If $|a|$ is the magnitude (or length) of vector a, $|s|$ is the magnitude of vector s, and θ is the angle subtended between these two vectors, then

$$a.s = |a||s|\cos\theta$$

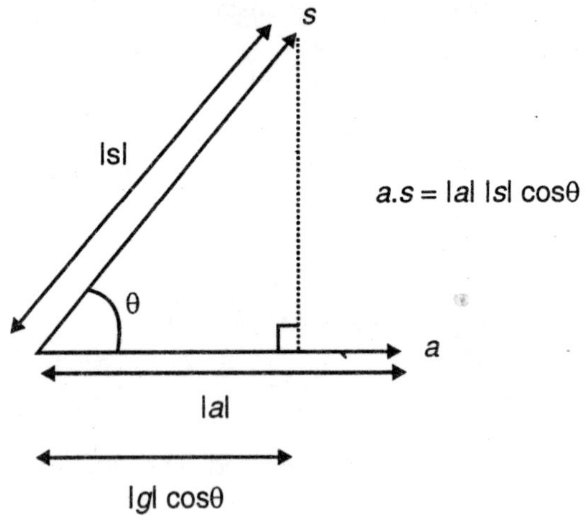

Fig. The scalar product

In other words, the scalar product of vectors a and s equals the product of the length of vector a times the length of that component of vector *s* which lies in the *same direction* as vector *a*. It immediately follows that if two vectors are mutually perpendicular (*i.e.* $\theta = 90°$) then their scalar product is zero. Furthermore, the scalar product of a vector with itself is simply the magnitude squared of that vector.

$$a \cdot a = |a|^2 = a^2$$

It is also apparent from equation that *a.s* = s .*a*, and a.(b + c) = a.b + a.c, and $a \cdot (\lambda s) = \lambda (a \cdot s)$

Incidentally, equation is obtained by taking the scalar product of equation with itself, taking the scalar product of equation with *a*, and then eliminating *t*.

PROJECTILE MOTION

As a simple illustration of the concepts introduced in the previous subsections, let us examine the following problem. Suppose that a projectile is launched upward from ground level, with speed v_0, making an angle θ with the horizontal. *Neglecting the effect of air resistance*, what is the subsequent trajectory of the projectile?

Our first task is to set up a suitable Cartesian coordinate system. The *x*-axis points vertically upwards (this is a standard convention), whereas the -axis points along the projectile's initial direction of horizontal motion. Furthermore, the origin of our coordinate system corresponds to the launch point. Thus, z = 0corresponds to ground level.

Neglecting air resistance, the projectile is subject to a constant acceleration g = 9.81 ms^{-1}, due to gravity, which is directed *vertically downwards*. Thus, the projectile's vector acceleration is written

$$a = (0, 0, -g)$$

Here, the minus sign indicates that the acceleration is in the minus -direction (*i.e.*, downwards), as opposed to the plus z-direction (*i.e.*, upwards).

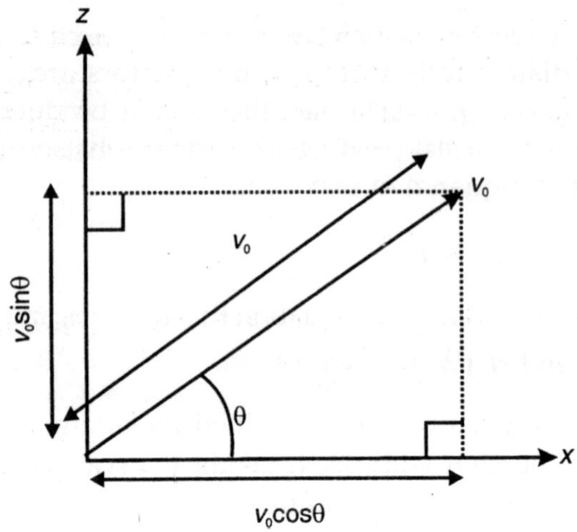

Fig. Coordinates for the projectile problem

What is the initial vector velocity v_0 with which the projectile is launched into the air at t = 0? Given that the magnitude of this velocity is v_0, its horizontal component is directed along the x-axis, and its direction subtends an θ angle with this axis, the components of v_0 take the form

$$v_0 = (v_0 \cos\theta, 0, v_0 \sin\theta)$$

Note that v_0 has zero component along the y-axis, which points *into* the paper.

Since the projectile moves with constant acceleration, its vector displacement $s = (x, y, z)$ from its launch point satisfies

Making use of equations, the x-, y-, and z-components of the equation are written

$$x = v_0 \cos\theta t$$

$$y = 0$$

$$z = v_0 \sin\theta t - \frac{1}{2} g t^2$$

respectively. Note that the projectile moves with *constant velocity*, $v_x = dx/dt = v_0 \cos\theta$, in the x-direction (*i.e.*, horizontally). This is hardly surprising, since there is zero component of the projectile's acceleration along the x-axis. Note, further, that since there is zero component of the projectile's acceleration along the y-axis, and the projectile's initial velocity also has zero component along this axis, the projectile never moves in the y-direction.

In other words, the projectile's trajectory is *2-dimensional,* lying entirely within the x - z- plane. Note, finally, that the projectile's vertical motion is entirely decoupled from its horizontal motion. In other words, the projectile's vertical motion is identical to that of a second projectile launched vertically upwards, at $t = 0$, with the initial velocity $v_0 \sin\theta$ (*i.e.,* the initial *vertical* velocity component of the first projectile)—both projectiles will reach the same maximum altitude at the same time, and will subsequently strike the ground simultaneously.

Equations can be rearranged to give

$$z = x\tan\theta - \frac{1}{2}\frac{gx^2}{v_0^2}\sec^2\theta$$

As was first pointed out by Galileo, this is the equation of a *parabola.* The horizontal range R of the projectile corresponds to its -coordinate when it strikes the ground (*i.e.,* when $z = 0$). It follows from the expression (neglecting the trivial result $x = 0$) that

$$R = \frac{2v_0^2}{g}\sin\theta\cos\theta = \frac{v_0^2}{g}\sin 2\theta$$

Note that the range attains its maximum value,

$$R_{max} = \frac{v_0^2}{g}$$

when $\theta = 45°$. In other words, neglecting air resistance, a projectile travels furthest when it is launched into the air at 45° to the horizontal.

The maximum altitude h of the projectile is attained when $v_z = dz/dt = 0$(*i.e.*, when the projectile has just stopped rising and is about to start falling). It follows from equation that the maximum altitude occurs at time $t_0 = v_0 \sin\theta / g$. Hence,

$$h = z(t_0) = \frac{v_0^2}{2g}\sin^2\theta$$

Obviously, the largest value of h,

$$h_{max} = \frac{v_0^2}{2g},$$

is obtained when the projectile is launched vertically upwards (*i.e.*, $\theta = 90°$).

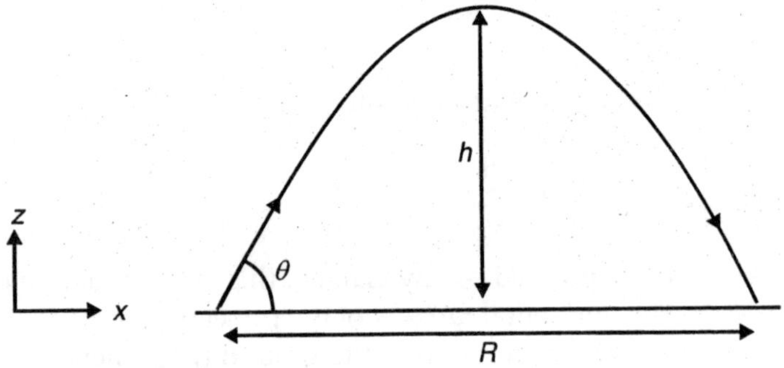

Fig. The parabolic trajectory of a projectile

RELATIVE VELOCITY

Suppose that, on a windy day, an airplane moves v_a with constant velocity with respect to the air, and that the air moves with constant velocity u with respect to the ground. What is the vector velocity v_g of the plane with respect to the ground? In principle, the answer to this question is very simple:

$$v_g = v_a + u$$

In other words, the velocity of the plane with respect to the ground is the vector sum of the plane's velocity relative to the air and the air's velocity relative to the ground. Note that, in general, v_g is parallel to neither v_a nor u. Let us now consider how we might implement equation in practice.

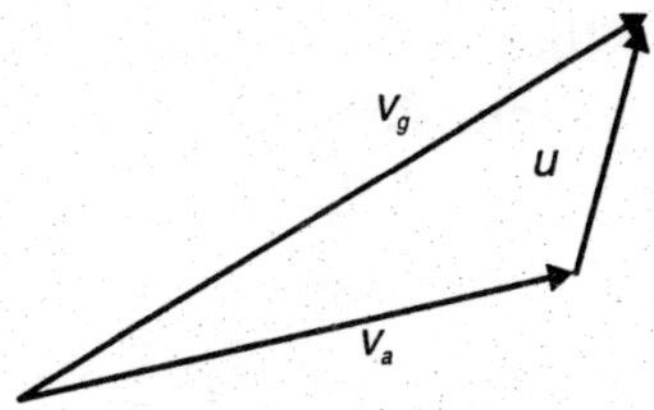

Fig. Relative velocity

As always, our first task is to set up a suitable Cartesian coordinate system. A convenient system for dealing with 2-dimensional motion parallel to the Earth's surface. The x-axis points northward, whereas the y-axis points eastward. In this coordinate system, it is conventional to specify a vector r in term of its magnitude, r, and its *compass bearing*, ϕ. A compass bearing is the angle subtended between the direction of a vector and the direction to the North pole: *i.e.*, the x-direction. By convention, compass bearings run from 0° to 360°. Furthermore, the compass bearings of North, East, South, and West are 0°, 90°, 180°, and 270°, respectively.

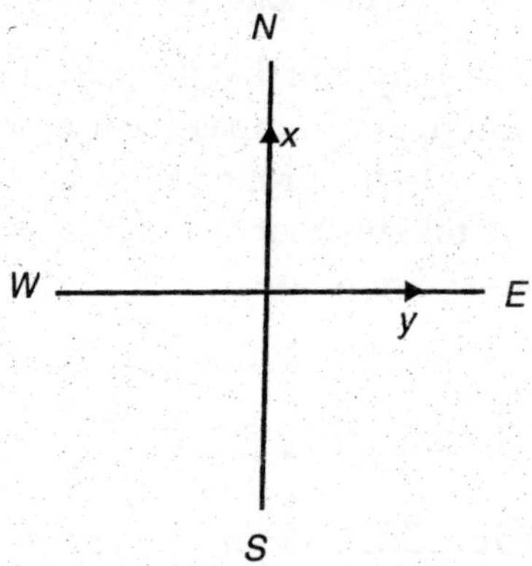

Fig. Coordinates for relative velocity problem

The components of a general vector r, whose magnitude is r and whose compass bearing is ϕ, are simply

$$r = (x, y) = (r\cos\phi, r\sin\phi)$$

Note that we have suppressed the z-component of r(which is zero), for ease of notation. Although, strictly speaking, figure only justifies the expression for ϕ in the range 0° to 90°, it turns out that this expression is generally valid: *i.e.*, it is valid for ϕ in the full range 0° to 360°.

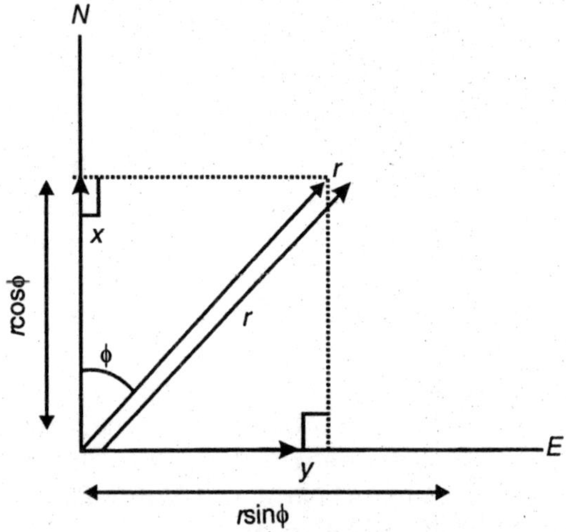

Fig. A compass bearing

As an illustration, suppose that the plane's velocity relative to the air is 300 km/h, at a compass bearing of 120°, and the air's velocity relative to the ground is 85 km/h, at a compass bearing of 225°. It follows that the components of v_a and u (measured in units of km/h) are

$$\mathrm{v_a} = (300\cos 120°, 300\sin 120°)$$

$$= (-1.500\times 10^2, 2.598\times 10^2)$$

$$\mathrm{u} = (85\cos 225°, 85\sin 225°)$$

$$= (-6.010\times 10^1, -6.010\times 10^1)$$

According to equation, the components of the plane's velocity v_g relative to the ground are simply the algebraic sums of the corresponding components of v_a and u. Hence,

$$\text{Vg} = \left(-1.500\times10^2 - 6.010\times10^1, 2.598\times10^2 - 6.010\times10^1\right)$$

$$= \left(-2.101\times10^2, 1.997\times10^2\right)$$

Our final task is to reconstruct the magnitude and compass bearing of vector v_g, given its components ϕ . The magnitude of v_g follows from Pythagoras' theorem:

$$v_g = \sqrt{\left(v_{gx}\right)^2 + \left(v_{gy}\right)^2}$$

$$= \sqrt{\left(-2.101\times10^2\right)^2 + \left(1.997\times10^2\right)}$$

$$= 289.9\text{km/h}$$

In principle, the compass bearing of v_g is given by the following formula:

$$\phi = \tan^{-1}\left(\frac{v_{gy}}{v_{gx}}\right)$$

This follows because $v_{gx} = v_g \cos\phi$ and $v_{gy} = v_g \sin\phi$. Unfortunately, the expression becomes a little difficult to interpret if v_{gx} is negative. An unambiguous pair of expressions for ϕ is given by :

$$\phi = \tan^{-1}\left(\frac{v_{gy}}{v_{gx}}\right),$$

if $v_{gx} \geq 0$; or

$$\phi = 180° - \tan^{-1}\left(\frac{v_{gy}}{\left|v_{gx}\right|}\right)$$

if $v_{gx} < 0$. These expressions can be derived from simple trigonometry. For the case in hand, equation is the relevant expression, hence

$$\phi = 180° - \tan^{-1}\left(\frac{1.997\times10^2}{2.101\times10^2}\right) = 136.5°$$

Thus, the plane's velocity relative to the ground is 289.9km/h at a compass bearing of 136.5°.

Problem : Major Applewhite receives the snap at the line of scrimmage, takes a seven step drop (*i.e.*, runs backwards 9 yards), but is then flushed out of the pocket by a blitzing linebacker. Major subsequently runs parallel to the line of scrimmage for 12 yards and then gets off a forward pass, 36 yards straight downfield, to Roy Williams, just prior to being creamed by the linebacker. What is the magnitude of the football's resultant displacement (in yards)?

Solution : As illustrated in the diagram, the resultant displacement of the football is the sum of vectors a, b, and c, which correspond to the seven step drop, the run parallel to the line of scrimmage, and the forward pass, respectively.

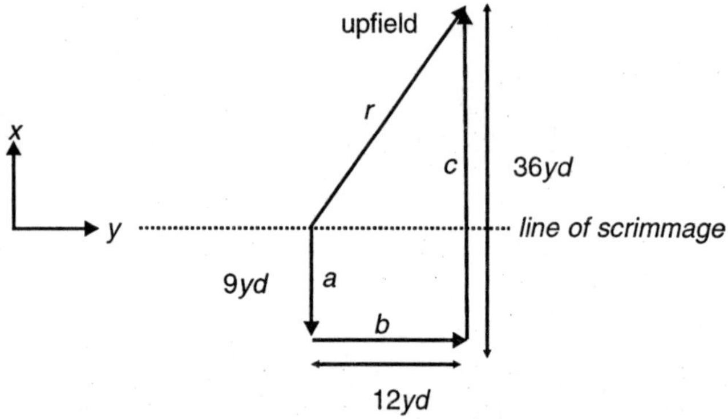

Using the coordinate system indicated in the diagram, the components of vectors, and (measured in yards) are

$$a = (-9, 0)$$

$$b = (0, 12)$$

$$c = (36, 0)$$

respectively. Hence the components of are given by

$$r = (x, y) = (-9+0+36, 0+12+0) = (27, 12)$$

It follows that the magnitude of the football's resultant displacement is

$$r = \sqrt{x^2 + y^2} = \sqrt{27^2 + 12^2} = 29.55yd$$

Problem : Legend has it that Gallileo tested out his newly developed theory of projectile motion by throwing weights from the top of the Leaning Tower of Pisa. Suppose that, one day, Gallileo simultaneously threw two equal weights off the tower from a height of 100m above the ground. Suppose, further, that he dropped the first weight straight down, whereas he threw the second weight horizontally with a velocity of 5m/s. Which weight struck the ground first? How long, after it was thrown, did it take to do this? Finally, what horizontal distance was traveled by the second weight before it hit the ground? Neglect the effect of air resistance.

Solution: Since both weights start off traveling with the same initial velocities in the vertical direction (*i.e.*, zero), and both accelerate vertically downwards at the same rate, it follows that both weights strike the ground simultaneously. The time of flight of each weight is simply the time taken to fall h = 100m, starting from rest, under the influence of gravity. Time is given by

$$t = \sqrt{\frac{2h}{g}} = \sqrt{\frac{2 \times 100}{9.81}} = 4.515s$$

The horizontal distance traveled by the second weight is simply the distance traveled by a body moving at a constant velocity u = 5m/s during the time taken by the weight to drop 100m. Thus,

$$R = ut = 5 \times 4.515 = 22.58\text{m}$$

Problem : A cannon placed on a 50m high cliff fires a cannonball over the edge of the cliff at v = 200m/s making an angle of $\theta = 30°$ to the horizontal. How long is the cannonball in the air? Neglect air resistance.

Solution : In order to answer this question we need only

consider the cannonball's vertical motion. At $t = 0$(*i.e.*, the time of firing) the cannonball's height off the ground is $z_0 = 50$m and its velocity component in the vertical direction is $v_0 = v\sin\theta = 200 \times \sin 30° = 100 m/s$. Moreover, the cannonball is accelerating vertically downwards at g = 9.91m/s². The equation of vertical motion of the cannonball is written

$$z = z_0 + v_0 t - \frac{1}{2} g t^2$$

where z is the cannonball's height off the ground at time t. The time of flight of the cannonball corresponds to the time t at which z = 0. In other words, the time of flight is the solution of the quadratic equation

$$0 = z_0 + v_0 t - \frac{1}{2} g t^2$$

Hence,

$$t = \frac{v_0 + \sqrt{v_0^2 + 2 g z_0}}{g} = 20.88 s$$

Here, we have neglected the unphysical negative root of our quadratic equation.

Problem : The Longhorns are down by 4 points with 5s left in the fourth quarter. Chris Simms launches a Hail Mary pass into the end-zone, 60 yards away, where B.J. Johnson is waiting to make the catch. Suppose that Chris throws the ball at 55 miles per hour. At what angle to the horizontal must the ball be launched in order for it to hit the receiver? Neglect the effect of air resistance.

Solution : The formula for the horizontal range of a projectile thrown with initial velocity v_0 at an angle θ to the horizontal is

$$: \quad R = \frac{v_0^2}{g} \sin 2\theta$$

In this case, $R = 60 \times 3 \times 0.3048 = 54.86 m$ and . Hence,

$$v_0 = 55 \times 5280 \times 0.3048 / 3600 = 24.59 m/s$$

$$\theta = \frac{1}{2}\sin^{-1}\left(\frac{R_g}{v_0^2}\right)$$

$$= \frac{1}{2}\sin^{-1}\left(\frac{54.86 \times 9.81}{(24.59)^2}\right) = 31.45°$$

Thus, the ball must be launched at 31.45° to the horizontal.

Problem : United Airlines flight UA589 from Chicago is 20miles due North of Austin's Bergstrom airport. Suppose that the plane is flying at 200mi/h relative to the air. Suppose, further, that there is a wind blowing due East at 60mi/h. Towards which compass bearing must the plane steer in order to land at the airport?

Solution : The problem in hand is illustrated in the diagram.

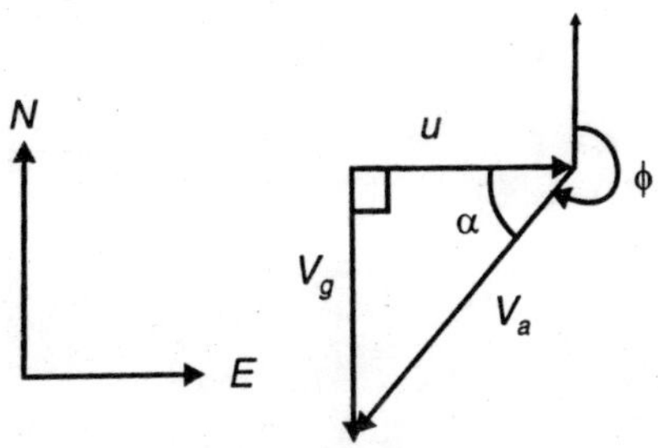

The plane's velocity v_g relative to the ground is the vector sum of its velocity v_a relative to the air, and the velocity u of the wind relative to the ground. We know that is directed due East, and we require v_g to be directed due South. We also know that $|v_a| = 200 mi/h$ and $|u| = 60 mi/h$. Now, from simple trigonometry,

$$\cos\alpha = \frac{|u|}{|v_a|} = \frac{60}{200} = 0.3$$

Hence,

$$\alpha = 72.54°$$

However, it is clear from the diagram that the compass bearing ϕ of the plane is given by

$$\phi = 270° - alpha = 270° - 72.54° = 197.46°.$$

Thus, in order to land at Bergstrom airport the plane must fly towards compass bearing 197.46°.

Chapter 8

Sound Waves

The science of mechanics grows out of the human sense of touch and sight by gradually extracting information, gathered through experimentation, into abstract theoretical frame work.

In the same way, the science of Sound grows out of the human sense of perception of hearing. Upon realization that these perceptions are manifestations of vibrations, the theory of sound can also be developed on the basis of principles of mechanics.

The elastic (longitudinal) waves in the air to which the human ear is sensitive are called sound waves. Thus 'sound' can be described as longitudinal elastic waves in air. (In fluids only longitudinal waves can be produced because fluids cannot sustain shear deformation. However on the free surfaces of liquids, transverse waves can also be produced.) These longitudinal sound waves are produced by vibrating objects, like vocal chords, various musical instruments, etc.

INTENSITY

Intensity of a sound wave is defined as Power flux i.e., amount of energy incident per unit normal area per unit time; i.e., Intensity I is measured in units of 1 Watt/m^2

The practical unit of Intensity is called 1 decibel ($1dB$) which really stands for comparison with some standard intensity (I_0) on the logarithmic (base 10) scale, and is defined as

$$\beta = 10\log_{10}\left(\frac{I}{I_0}\right)^{dB}$$

If $I_0 = 10^{-12}$ watt/m^2, the threshold of hearing i.e. the intensity that is just audible to a normal ear then b = 60 *dB* for normal conversation.

$$\text{Thus, } 60 = 10\log_{10}\left(\frac{I}{I_0}\right)$$

$$\therefore \log_{10}\left(\frac{I}{I_0}\right) = 6$$

$$\therefore \frac{I}{I_0} = 10^6$$

$$\text{or } I = 10^6\, I_0$$

Thus the intensity of sound in normal conversation is a million times greater than the just audible intensity of sound.

Loudness

Loudness of sound is the qualitative or subjective measure of intensity and generally corresponds to intensity. A sound of a particular intensity with different frequencies can be perceived as having different loudness by the same person; or two persons may perceive the loudness to be different even if they both listen to the sound of the same intensity.

PITCH

The response of the human ear to sound in terms of shrillness of sound is called pitch. This is against quantitative and subjective measure, and it corresponds to the objective measure called frequency. Thus a high pitch or shriller sounds correspond to a high frequency etc.

DOPPLER EFFECT

The apparent change in frequency of a sound wave, detected by an observer, in relative motion with respect to the source of sound waves, is called Doppler effect.

For elastic or mechanical waves such as sound waves, in some material medium, the effect depends on whether the listener is moving towards the stationary source or the source is moving towards the stationary listener. This can be seen even though the relative velocity of motion between the listener and the source is the same.

Source (*S*) in motion and Listener (*L*) at rest, relative to some fixed interval frame of reference..asp.asp.asp.asp.asp.asp.asp.asp

(a) Source (*S*) moving towards the stationary listener (*L*).

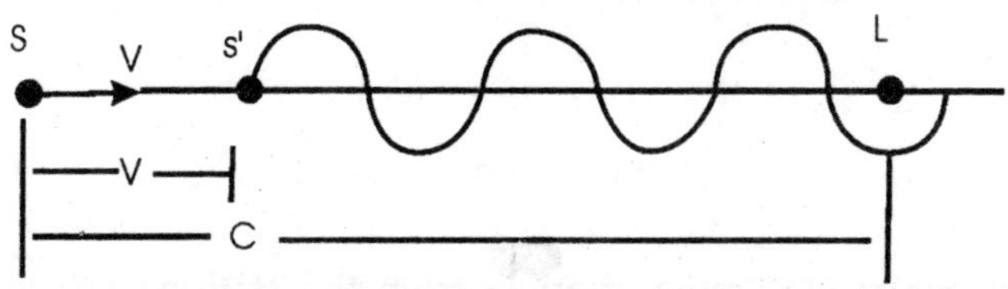

Let n : Frequency of Sound waves emitted by 'Source' S

n′ : Apparent frequency of these wave received by listener *L*

S : initial position of 'source' at $t = 0$ sec

S′ : final position of 'source' at $t = 1$ sec

v : velocity of approach of *S* to *L* along the line joining them.

c : velocity of sound in air (Still)

For convenience, we choose initial distance between *S* and *L* to be numerically equal to '*C*', so that if *S* were stationary then all *n* waves emitted by source in 1 sec. will be received by *L* within interval of 1 sec.

To the listener all the n waves emitted by the source appear to be compressed in the distance $S'L = C - V$,

$$\therefore \lambda' = \text{Apparent wave length}$$

$$= \frac{C-v}{n}$$

$$\therefore \frac{C}{n'} = \lambda'$$

$$= \frac{C-v}{n}$$

$$\therefore n' = n\left(\frac{C}{C-v}\right)$$

$$= \frac{n}{1-\frac{v}{C}}$$

$$n' = \frac{n}{1-\frac{v}{C}} \quad \text{i.e. } n' > n$$

(b) Source (S) moving away from the stationary listener (L)

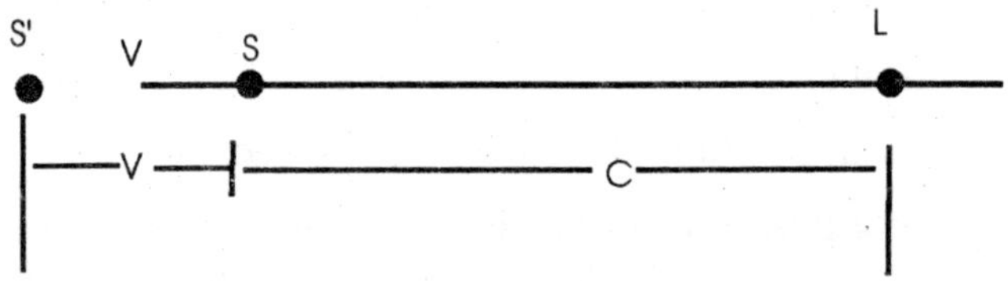

This is similar to C as but $\vec{v} \equiv -\vec{v}$

$$\therefore n' = \frac{n}{1+\frac{v}{C}}$$

Source (S) stationary and listener (L) in motion, relative to some fixed reference frame (inertial)

(a) Listener (L) moving towards the stationary Source (S)

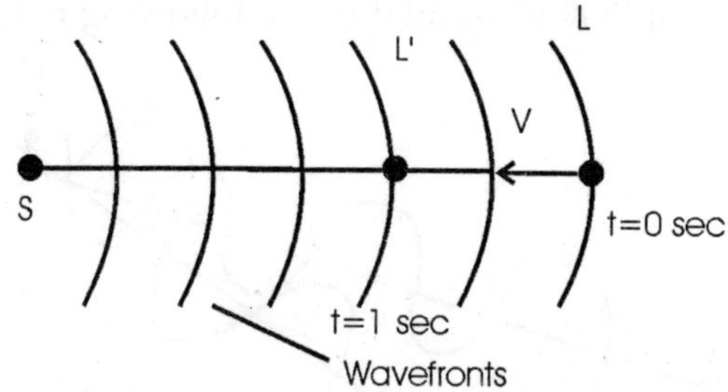

If *L* were stationary then only x waves will be received in 1 sec, but as *L* is approaching *S*, the extra number of waves within distance $V = LL'$ will be received therefore, the number of waves received by *L* in 1 sec. = n + number of waves in distance v.

$$\text{i.e, } n' = n + \left(\frac{v}{c}\right)n$$

$$n' = n\left(1 + \frac{v}{c}\right) \quad \text{i.e, } n' > n$$

(b) Listener (*L*) moving away from the stationary source (*S*)

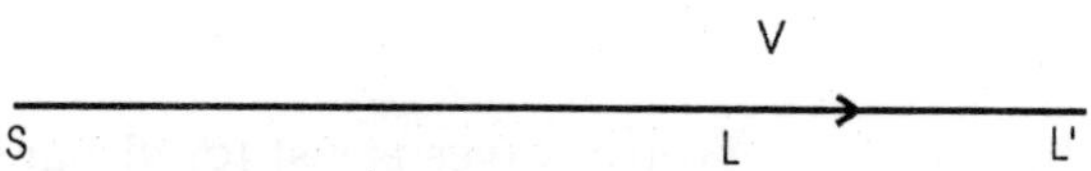

Compared to (a) above $\vec{v} = -\vec{v}$

$$n' = n\left(1 - \frac{v}{c}\right) \text{ i.e } n' < n$$

SUPERPOSITION PRINCIPLE

The effect of two or more waves, on the oscillations of particles of a medium at which the waves arrive simultaneously, is given by the superposition principle.

It states that "The resultant vibration of any particle at which two or more waves arrive simultaneously is given by the Vector addition of displacements given to the particle of each wave independently of the other."

The principle is illustrated in the following diagram:

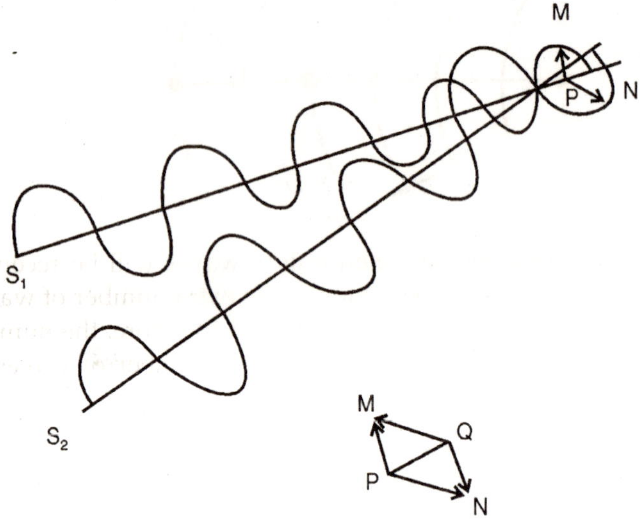

$\vec{PM}$: Displacement of P due to waves from S_1

$\vec{PN}$: Displacement of P due to waves from S_2

$\vec{PQ} = \vec{PM} + \vec{PN}$: Result displacement of P.

Note

1. The use of transverse waves is just for visual effect.
2. If the angle subtended at *P* by the lines of propagations S_1P and S_2P of the waves is negligible, i.e. the distance between sources S_1S_2 is much smaller than the distance of either of the sources from '*P*',then *PM* and *PN* are either parallel to each other or are anti - Parallel to each other.
3. When $\vec{PM} \| \text{to} \vec{PN}$ then $PQ = PM + PN$ and in same direction as PM or PN. This condition is called constructive interference.
4. When *PM* anti- || to PN then PQ = PM–PN or PQ = *PN- PM* which ever is +ve and in the direction of the greater of the two.

THE PHENOMENON OF BEATS OF SOUND

When two vibrating objects (Sources of audible Sound) vibrate with slightly different frequencies at a place far away - compared to the separation between the sources - from the sources the loudness (intensity) of sound waves and waves periodically, this phenomenon is called Beats of Sound. The beat frequency i.e., frequency of waxing and waning of sound is found to be equal to the positive difference between the two interfering frequencies.

The equations of two waves from the two sources with the same amplitude of waves; with sources very close to each other and to x- axis can be written as

$$Y_1 = r\sin 2\pi\left(n_1 t - \frac{x}{v}\right)$$

$$\text{and } Y_2 = r\sin 2\pi\left(n_2 t - \frac{x}{v}\right)$$

Since the waves are traveling along the same direction i.e., x- axis, the superposition principle gives the resultant expression as,

$$Y = Y_1 + Y_2$$

$$Y = r\left[\sin 2\pi\left(n_1 t - \frac{x}{2}\right) + \sin 2\pi\left(n_2 t - \frac{x}{2}\right)\right]$$

$$Y = r\left[\sin 2\pi n_1 t + \sin 2\pi n_2 t\right]$$

[shifting the origin from sources to the point at distance 'x' does not alter the characteristics of interfering waves]

$$Y = \left[2r\cos\left\{2\pi\left(\frac{n_1 - n_2}{2}\right)t\right\}\right]$$

$$\sin\left[2\pi\left(\frac{n_1 + n_2}{2}\right)^t\right]$$

where $A = 2r\left\{\cos 2\pi\left(\frac{n_1 - n_2}{2}\right)\right\}$ is the resultant amplitude of vibration at P, and $n = \frac{n_1 + n_2 i}{2}$ is the frequency with which the particle at P vibrates.

Now, waxing and waning correspondent to maximum and minimum intentity respectively, i.e. on

$$I \alpha A^2 \alpha \cos^2\left\{2\pi\left(\frac{n_1 - n_2}{2}\right)t\right\}$$

$$\therefore \text{when } 2\pi\left(\frac{n_1 - n_2}{2}\right)t = m\pi, m = 0,$$

then waxings will occur and,

$$\text{when } 2\pi\left(\frac{n_1 - n_2}{2}\right)t = (2K+1)\frac{\pi}{2}, K = 0,1,2,...$$

then warnings will occur.

Therefore frequency of waxings or wanings is $N = | n_1 - n_2 |$

The Phenomenon of Beats can be illustrated graphically also, as shown below:

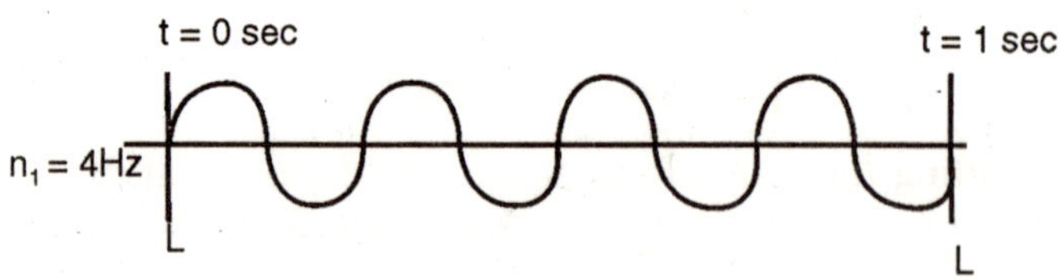

The above diagrams shows the number of waves arriving from the two sources of frequencies n_1 and n_2 at the position of the Listener over the interval of 1 sec. The diagram (c) represents the effect of these superposed waves resulting into 4 maximas (waxings) and minimas (wanings) indicating that beat frequency is $N = n_2 - n_1 = |n_1 - n_2|$ in general. These diagrams are only illustrative ones, since n_1, $n_2 < 20$ Hz and N < 20 Hz, the sounds are inaudible to humans.

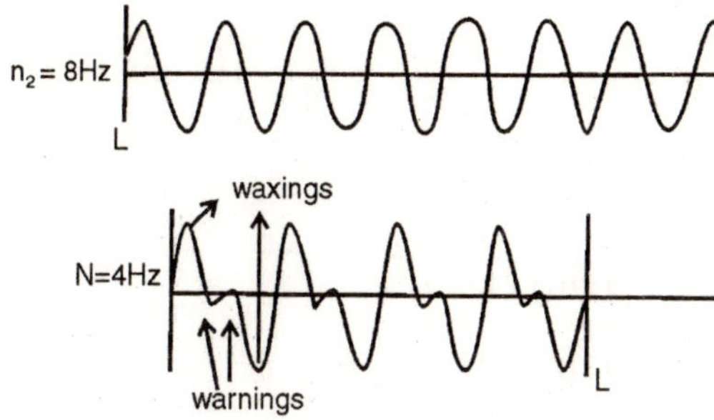

STANDING OR STATIONARY WAVES

If a taut string that is held fixed at its ends is suitably plucked, then the visual pattern of a loop is observed to be formed between the fixed ends. This and similar phenomena are due to the formation of standing waves. The standing waves are formed due to superposition of two waves having the same characteristics but traveling in opposite directions along same path.

The analysis of standing waves can be obtained from superposition of two traveling harmonic waves. This is as follows:

The equations of the waves traveling along +ve and -ve X- axis having same amplitudes, wave lengths frequencies and velocities can be written as

$$Y_1 = r \sin 2\pi\left(\frac{t}{T} - \frac{x}{\lambda}\right)$$

and $$Y_2 = r \sin 2\pi\left(\frac{t}{T} + \frac{x}{\lambda}\right)$$

$\therefore$ The Resultant wave equation according to superposition principle is,

$$Y = Y_1 + Y_2$$

$$= 2r \sin 2\pi\left(\frac{t}{T}\right) \cos 2\pi\left(\frac{-x}{\lambda}\right)$$

$$= 2r\cos\left(\frac{2\pi x}{\lambda}\right)\sin(2\pi nt)$$

$$(\because \cos(-\theta) = \cos\theta)$$

where $x = \frac{t}{T}$

$$= A\sin(2\pi nt)$$

$$= A\sin(\omega t)$$

where A= 2r cos $\left(\frac{2\pi x}{\lambda}\right)$ is the resultant amplitude and $\omega = 2\pi x = \frac{2\pi i}{T}$ is the angular frequency of vibration of each particle.

The amplitude A = 2r cos $\left(\frac{2\pi x}{\lambda}\right)$ is position dependent, the formation of Nodes-points at which no vibrations- and anti- Nodes- points at which vibrations with maximum amplitudes occur in standing wave. The Equation Y = A sin (wt); A= 2r cos $\left(\frac{2\pi x}{\lambda}\right)$ is called the equations of the standing wave.

Nodes	**Antinodes**
$A = 0(\text{by definition})$	$A = \pm 2r$
$\therefore \cos 2\pi x = 0$	$\therefore \cos 2\pi x = \pm 2r$
$\therefore \frac{2\pi x}{\lambda} = (2k+1)\frac{\pi}{2}$	$\frac{2\pi x}{\lambda} = m\pi,$
k = 0,1,2,3.......	m = 0,1,2,3.......
$x_k = (2k+1)\frac{\lambda}{4}$	$x_m = m\frac{\lambda}{2}$
$i.e.\ x = \frac{\lambda}{4}, \frac{3\lambda}{4}, \frac{5\lambda}{4}...$	$i.e. x = \frac{\lambda}{4}, \frac{2\lambda}{4}, \frac{3\lambda}{4}...$

It is obvious that between any two successive nodes, an anti-node and conversely between any two successive anti-nodes a node, occurs exactly in the middle. The nodes and anti-nodes are formed alternately with separation of $\lambda / 4$ and the distance between successive nodes and anti-nodes is $\lambda / 2$.

The formation of standing waves and nodes, anti-nodes is illustrated in the following diagram.

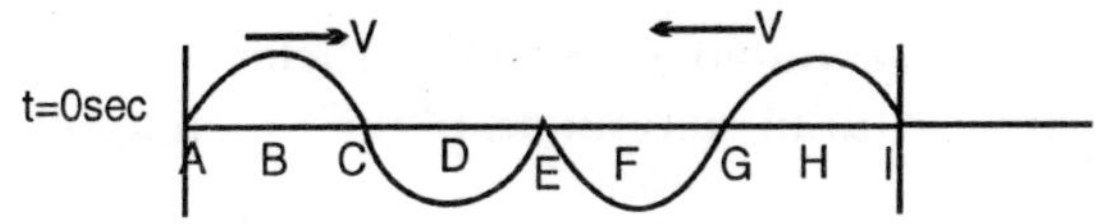

Figure 7 (a)

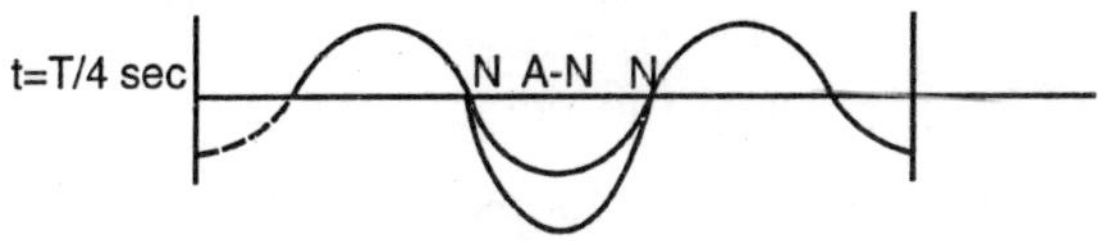

Figure 7 (b)

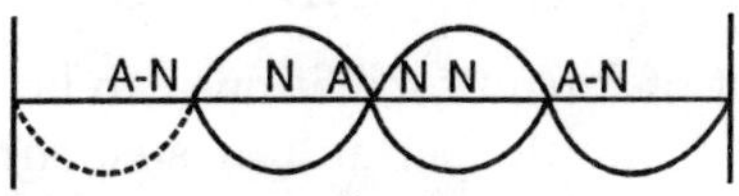

Figure 7(c)

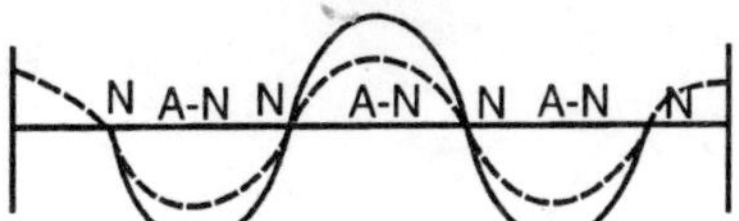

Figure 7 (d)

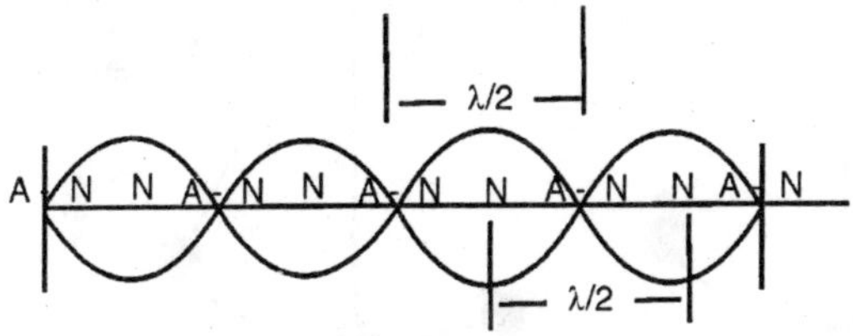

The diagrams (*a*) to (*e*) represent the waves at successive instants as they propagate in their respective directions. The dotted curves represent the generations of further pulses, as the particles at A & I are allowed to vibrate further over a of time period T.

As the waves start superposing over each other at $\frac{T}{4}$ sec, the particles at E, C and G, A and I vibrate with maximum aplitude and hence anti-nodes (A-N) occur at these positions; whereas particles at D and F, B and G do not vibrate at all. These positions are called Nodes.

It should be noted that the particles where A-N are formed virate with maximum amplitude therefore for waves of frequency $\geq$ 10 Hz i.e. $T \leq \frac{1}{10}$ sec; these particles move from the maximum to -ve maximum along Y-axis in an interval of $\geq \frac{1}{20}$, hence the persistence of vision gives rise to appearance of stationary loops when the standing waves are formed on strings.

FORCED OSCILLATIONS AND RESONANCE

The concept of forced oscillations can be explained with an example of a simple pendulum suspended from rigid support and two pendulums suspended side by side from a rubber string.

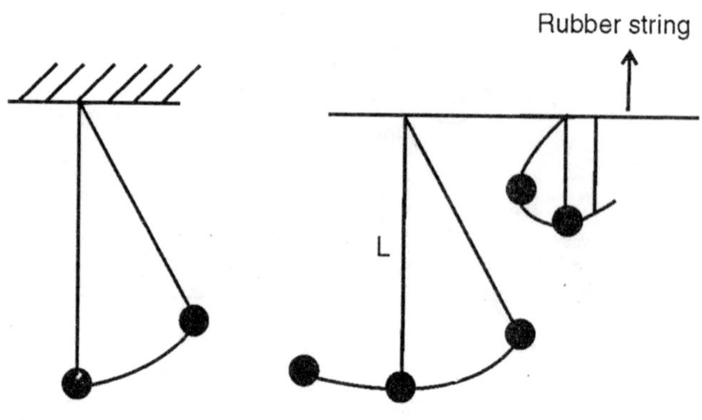

The concept of forced oscillations can be explained with an example of a simple pendulum suspended from a rigid support and two pendulums suspended side by side from a rubber string.

The Oscillations of pendulum of length L are called free oscillations. Similarly oscillations of pendulum of fixed length L are also called free oscillations. The pendulum of length 'l' (variable) due to free oscillations of pendulum of length L are called forced oscillations. The forced oscillations are communicated through a medium (Rubber String in this example); but non- mechanical waves such as Electro-Magnetic waves can be communicated through empty space as well as through the electro-magnetic field.

The response of pendulum of length 'l' in terms of its amplitude compared with fixed amplitude of the freely oscillating pendulum varies with the length 'l' and it is found that the pendulumperforming forced oscillationsoscillateswith maximumamplitudewhenits lengthismadeequal to L.

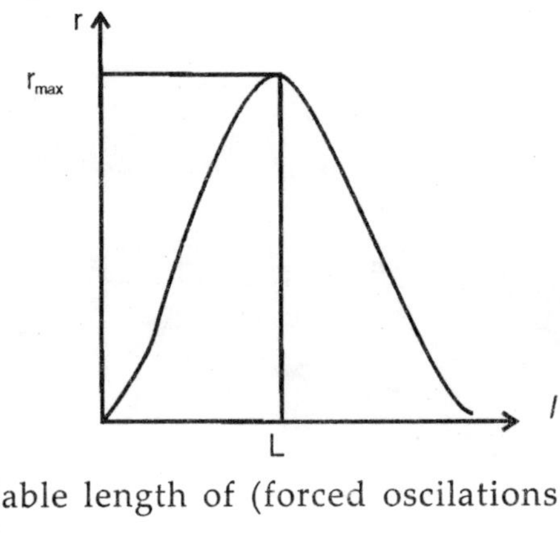

l: Variable length of (forced oscilations performing pendulum)

r: its amplitude

At l = L, i.e, when frequency (natural) 'n' of pendulum of length 'l' is forced upon it by pendulum of length 'L' then the first pendulum has maximum amplitude of forced oscillations; and since energy of oscillations is proportional to r^2, therefore maximum energy is absorbed when l = L from the same constant supply of energy by freely oscillating pendulum. The phenomenon of maximum absorption of energy by forced system is called resonance phenomenon.

The maximum absorption of energy by forced system when its natural frequency is forced on it is the phenomenon of resonance.

The Phenomenon of resonance in the basic principle in

1. Tuning of musical instruments.
2. Tuning of radio sets.
3. Tuning of TV sets.
4. Warning for army troops to walk over and not to march over road bridges.
5. Similar warning to vehicles to drive within maximum permissible speed limit over road bridges.
6. Radar imaging and so on.

Chapter 9

Mechanics of Fluids

Newtonian Mechanics can be extended to any macroscopic system of particles except when the gravitational field is not very strong, and therefore the behaviour of liquids and gases (fluids) should be amenable to analysis in terms of Newtonian Mechanics.

HYDROSTATIC PRESSURE IN A FLUID

Pressure in general is defined as force per unit normal area. To derive the expression for hydrostatic pressure, i.e., pressure when the fluid as a bulk system is at rest, consider an imaginary cylinder of fluid in the suspended state in the fluid itself.

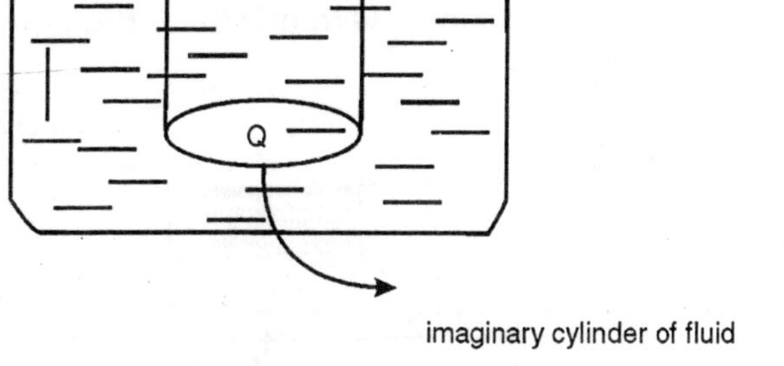

Let 'A'= Area of Cross section and 'h'=height of cylinder,

Let P_1=Pressure of the fluid at the top face P &

P_2= Pressure of the fluid at the bottom face Q

Then F_2= Upward force on the cylinder due to P_2

F_1=Downward force on the cylinder due to P_1

W= Weight of the cylinder.

For equilibrium of the cylinder

Total downward force = Total upward force

$$\therefore W + P_1 A = P_2 A$$

$$\therefore (P_2 - P_1) A = W$$

but W = Mass of fluid in cylinder×g

= Volume of fluid in cylinder×ρ×g

= Ah×ρ×g

$$\therefore (P_2 - P_1) A = Ah\rho g$$ (Cancelling A from both side we get,)

but $h = h_2 - h_1$

$$\therefore (P_2 - P_1) = (h_2 - h_1)\rho g$$

∴ if $h_1 = 0$ i.e. the free surface where $P_1 = 0$ then $P_2 = h_2 \rho g$

∴ at any depth 'h' below the free surface of the fluid

$$P = h\rho g$$

PASCAL'S LAW

The ratio of forces is in the same proportion as the area of the cross-section when the forces is transmitted through a fluid at rest.

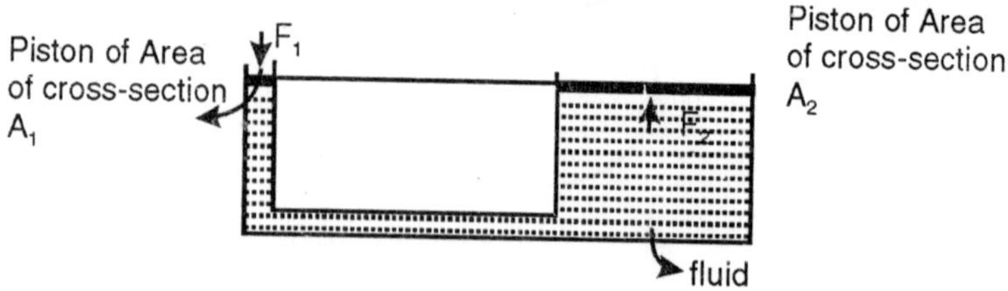

Since the two pistonsare on the same horizontal level floating above the fluid,

$$\therefore p_1 = p_2$$

$$\therefore \frac{F_1}{A_1} = \frac{F_2}{A_2}$$

$$\therefore \frac{F_2}{F_1} = \frac{A_2}{A_1}$$

$$or\ F_2 = \frac{A_2}{A_1} F$$

Thus depending on $\frac{A_2}{A_1}$ the force F_1 becomes filed by the factor $\frac{A_2}{A_1}$

ARCHIMEDES PRINCIPLE

When an object is immersed wholly or partly in a fluid then the apparent loss in weight of the fluid displaced by the immersed part of the object.

If W_1= Wt. of the object in air

W_2= Wt. of the object immeresed in a fluid

ρ = Density of the fluid

V = Volume of the fluid displaced by the immersed part of the object (obviously equals the volume of the immeresed part of the object.)

Then,

$W_1 = W_2$=Loss in weight (apparent)

$V\dot{\rho}g$ = Weight of the fluid displaced by the immersed part of the object

$$\therefore W_1 - W_2 = V\dot{\rho}g$$

Note: W_1 in principle should also be considered as

$W_1 = mg - V\rho_{air}\, g$;

where *m*: Mass of the object, *V*: Volume of the object but ρ_{air} is 3 useless of a magnitude much smaller than ρ water and hence the effect of ρ_{air} is negligible for all practical purposes.

This principle was discovered by Archimedes long before Newtonian Mechanics was developed It is an empricial law based on observations. Of course it can be derived as a consequence of equilibrium analysed through forces acting on the object. For simplicity, considera solod cylinder of metal suspended from a spring balance and immersed in a liquid as shown.

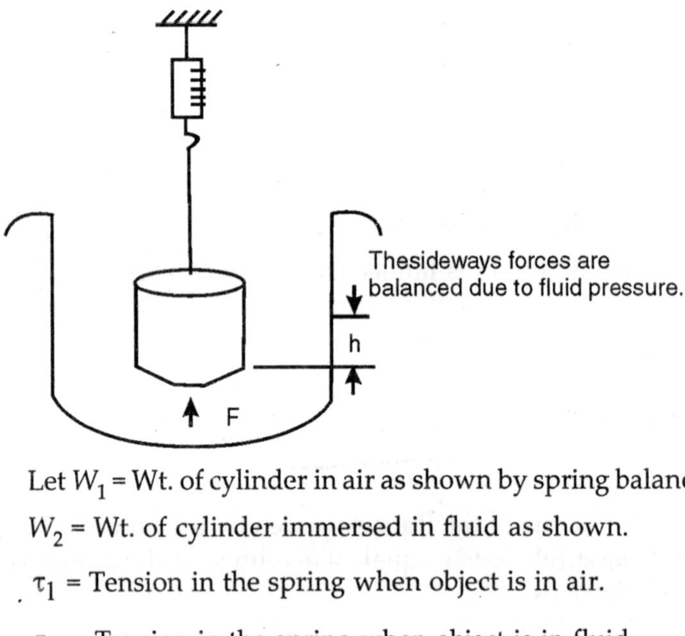

Let W_1 = Wt. of cylinder in air as shown by spring balance.

W_2 = Wt. of cylinder immersed in fluid as shown.

τ_1 = Tension in the spring when object is in air.

τ_2 = Tension in the spring when object is in fluid.

Then $W_1 = \tau_1$

& $W_2 = \tau_2$

aslo for equilibrium in immersed condition

$$\tau_2 + F = W_1$$

$$\therefore W_1 - \tau_2 = F$$

$$\therefore W_1 - W_2 = F$$

but $F = PA$ (A: Area of cross- section of the cylinder)

$$= h\rho g A$$

$$= (Ah)\rho A$$

$$= V\dot{\rho}g$$

$$\therefore W_1 - W_2 = V\dot{\rho}g$$

Corollary: (1) If $W_2 = 0$ then the object is either in a submerged or floating condition

$$W_1 = V\dot{\rho}g, \text{but } V \leq (V : \text{Vol. of object})$$

$$\therefore Vdg = V\dot{\rho}g\,(d : \text{density of object})$$

$$\therefore \cancel{V}\,d\cancel{g} \leq \cancel{V}\dot{\rho}\cancel{g}$$

$$\therefore d \geq P$$

thus, if $d < \rho$ object is floating

$d = \rho$ object is submerged

(2) If $W_1 > V\rho g$, the object sinks

i.e., $\cancel{Vdg} > \cancel{V\rho g}$

i.e., $d > \rho$

$$\text{Hence } d \begin{cases} > \rho \rightarrow \text{object sinks} \\ = \rho \rightarrow \text{object is submerged} \\ < \rho \rightarrow \text{object floats} \end{cases}$$

MOVING FLUIDS AND BERNOULLI'S EQUATION

The motion or flow of a fluid is a complex phenomenon, which in principle can be analysed in terms of Newtonian Mechanics, as applied to a system of particles.

Two types of flow of fluids can be characterized. They are the streamlined flow and the turbulent flow.

The flow of water in naturally occurring streams is an example of steady-streamlined flow whereas the flow of water in rapids or waterfalls is the turbulent flow.

The steady-stream lined flow is characterized by stream lines and tubes of stream lines. When a non-viscous fluid flows through a certain region of space such that at a given point the velocity of all fluid particles that flow past that point is same at all times (though at different points the velocity of fluid may be different but again it will be independent of time), then such a flow is called the steady - stream lined flow.

The line (curve) connecting such points is called a stream line. Stream lines have the following properties:

1. Only one stream line can pass through any point.
2. A tangent at any point to the stream line represents the path of motion of a fluid particle passing through that point.

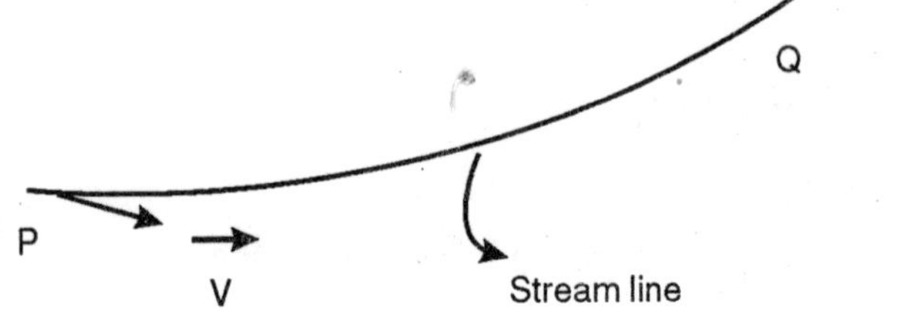

The diagram above represents a stream line. Any particle at P will have the same velocity V and will follow the same path from P to Q. Also particles that arrive at Q will have same velocity V at all times.

TUBE OF FLOW

Consider any area of Cross-section A in a fluid in a steady flow. Consider the stream lines of flow from each point on the boundary of cross-section, these stream lines enclose a tube; such a tube is called a tube of flow. The properties of tubes of flow are:

1. Fluid in different tubes cannot mix with each other.
2. The same fluid particles will keep on moving through a given tube of flow.

EQUATION OF CONTINUITY

The conservation of the mass of fluid in a tube of flow, immediately results into the equation of continuity.

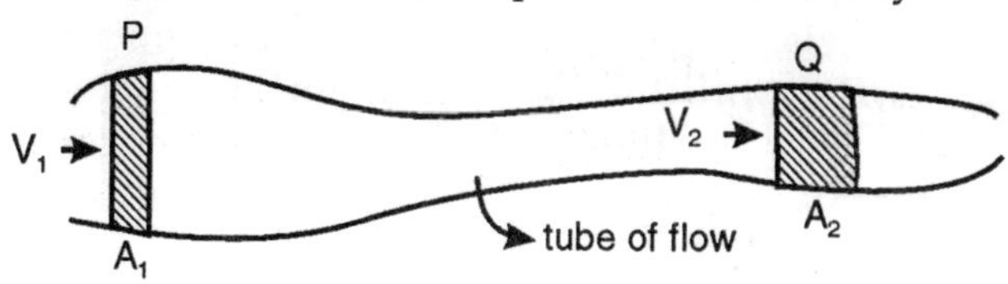

Consider the steady - Stream lined flow of a non viscous fluid; consider a tube of flow of such a fluid in motion as shown. Let *P* and *Q* be two points and let A_1 and A_2 be areas of cross-sections of the tube of flow normal to the velocities v_1 and v_2 respectively at *P* and *Q*.

Consider the shaded portions of the fluids at *P* and *Q*. Then, in a given time interval, $(\Delta m)_1$ = mass of fluid at *P* crossing A_1 and $(\Delta m)_1$ = mass of fluid at *Q* crossing A_2 must be equal, because the same fluid particles keep on moving in a given tube of flow.

$$\text{i.e, } (\Delta m)_1 = (\Delta m)_2$$

$$\text{But, } (\Delta m)_1 = (\Delta v)_1 \times \rho_1$$

$$= A_1 \, v_1 \Delta v \rho_1$$

where $(\Delta v)_1$ and $(\Delta v)_2$ are volumes of fluids in shaded parts at *P* and *Q* of respectively and $\rho_1 \rho_2$ are density of fluid at *P* and *Q*.

$$\text{and } (\Delta m)_2 = (\Delta v)_2 \times \rho_2$$

$$= A_2 V_2 \Delta t \times \rho_2$$

$$\therefore A_1 V_1 \cancel{\Delta t} \times \rho_1 = A_2 V_2 \rho_2 \cancel{\Delta t}$$

$$A_1 V_1 \rho_1 = A_2 V_2 \rho_2$$

If the fluid is incompressible, i.e. $\rho_1 = \rho_2$

$$A_1 V_1 = A_2 V_2$$

$$\text{i.e., } AV = \text{constant}$$

$$\text{Now } AV = A \times \frac{\text{distance}}{\text{time}}$$

$$= \frac{\text{volume}}{\text{time}}$$

$$= \text{rate of flow}$$

$$\therefore \text{Rate of flow} = AV = \text{constant}$$

Bernoulli's Equation

The equation of continuity as explained above is a consequence of conservation of mass; whereas Bernoulli's equation is consequence of the work - energy theorem or conservation of energy.

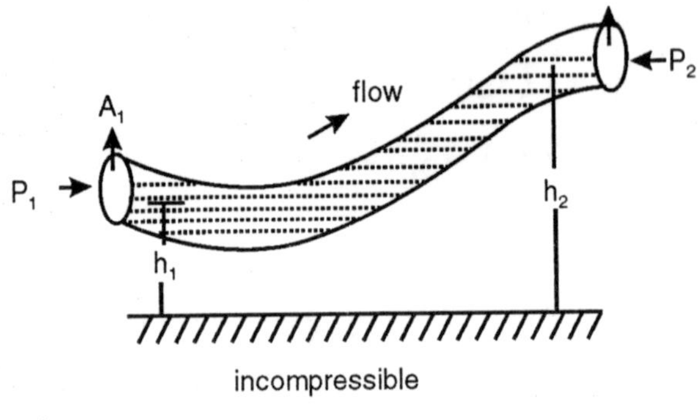

Consider an incompressible non-viscous, non-rotational, steady-streamlined flow of a fluid through a pipe (equivalent to a tube of flow) as shown.

Then = ΔW Total work done on fluid between C and D

$$P_1 A_1 V_1 \Delta t = P_2 A_2 V_2 \Delta t$$

but
$$(\Delta m)_1 = \rho_1 A_1 V_1 \Delta t$$
$$= (\Delta m)_2$$
$$= \rho_2 A_2 V_2 \Delta t$$

$$\therefore \Delta W = \frac{P_1 (\Delta m)_1}{\rho_1} - \frac{P_2 (\Delta m)_2}{\rho_2}$$

$$= P_1 \frac{\Delta m}{\rho} - P_2 \frac{\Delta m}{\rho}$$

$$\left[(\Delta m)_1 = (\Delta m)_2 \; \rho_1 = \rho_2\right]$$

$$= (P_1 - p_2)\frac{\Delta m}{\rho}$$

and E = Total gain in energy
= gain in P.E + gain in K.E.

$$= \Delta m g(h_2 - h_1) + \frac{1}{2}(\Delta m) v_2 2 - \frac{1}{2}(\Delta m) v_1 2$$

Now, from work-energy theorem

$$\Delta w = \Delta E$$

$$\therefore \left(\frac{P_1}{\rho}\right)\Delta m - \left(\frac{P_2}{\rho}\right)\Delta m = \left[\Delta m\, g h_2 - g h_1 + \frac{1}{2}V_2^2 - \frac{1}{2}V_1^2\right]$$

$$(P_1 - P_2)\frac{\Delta m}{\rho} = \left[\Delta m\, g h_2 - g h_1 + \frac{1}{2}V_2^2 - \frac{1}{2}V_1^2\right]$$

$$\therefore P_1 - P_2 = \rho g h_2 + \frac{1}{2}\rho v_2^2 - \rho g h_1 - \frac{1}{2}\rho v_1$$

$$\therefore \quad P + \rho g h_1 + \frac{1}{2}\rho v_1^2 = P_2 + \rho g h_2 + \frac{1}{2}\rho v_2^2$$

$$\therefore P + \rho g h + \frac{1}{2}\rho v^2 = \text{constant}$$

This equation is called bernouli's equation

Application of Equation of Continuity and Bernoulli's Equation

Ventury meter:

It is a device used for measuring the rate of flow of a non-viscous, incompressible fluid in non-rotational and steady-stream lined flow.

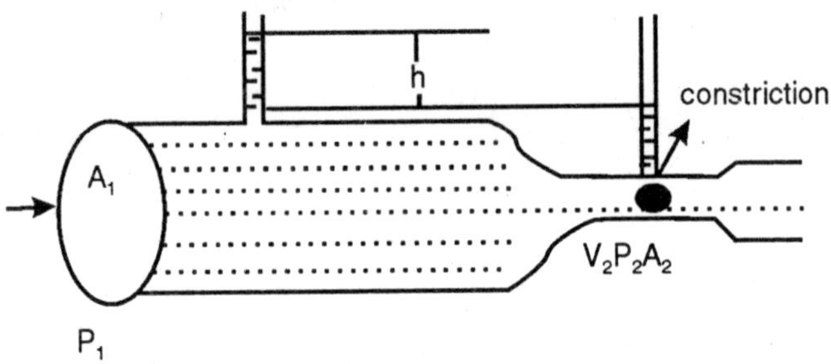

Let the fluid enter with velocity v_1 and at the left end. Let A_1 be the area of cross section and P_1 be the liquid pressure. Let it flow out of a constriction with v_2, and let P_2, A_2 be the pressure and area of the cross section at the c $A_1v_1 = A_2v_2$ onstriction.

Then $A_1v_1 = A_2v_2$ -------(1) from Equation 4 of contijnuity

and $P_1 + \rho g_1 h_1 + \frac{1}{2}\rho v_1^2 = P_2 + \rho g_2 h_2 + \frac{1}{2}\rho v_2^2$

-----(2) from Bernouli's equation.

but $h_1 = h_2$

$$\therefore (P_1 - P_2) = \frac{1}{2}\rho\left(v_2^2 - v_1^2\right)$$

then $P_1 - P_2 = h\rho g$

$$\therefore \frac{1}{2}\rho\left(v_2^2 - v_1^2\right) = h\rho g$$

$$\therefore v_2^2 - v_1^2 = 2h\rho g$$

$$\text{then, } v_2 = \frac{A_1 v_1}{A_2}$$

$$\frac{A_1}{A_2} v_1^2 - v_1^2 = 2gh$$

$$v_1 = \left[\frac{2gh}{\left(\frac{A_1}{A_2}\right)^2 - 1}\right]^{1/2}$$

$$\text{and } v_2 = \frac{A_1}{A_2}\left[\frac{2gh}{\left(\frac{A_1}{A_2}\right)^2 - 1}\right]^{1/2}$$

Dynamical Lift

Consider a ball in spin motion about a horizontal axis and let the ball also have translational motion in horizontal direction but perpendicular to the axis of spin motion moving in otherwise still air.

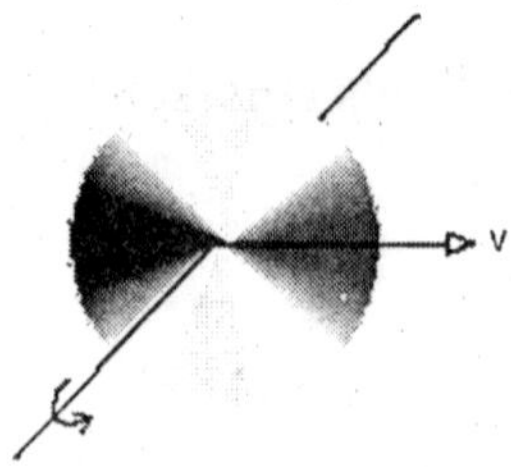

If we consider the frame of reference in which the centre of the ball is at rest then air will flow and pass the ball in the opposite direction.

Now at 'H' velocity of air is increased by the spin of the ball and at 'L' velocity of air is decreased by the spin of the ball.

Therefore pressure at H is decreased and is increased at L according to Bernoulli's equation. As a result net force F acts on the ball from the direction of L to H direction, i.e., the ball is lifted vertically upwards. The lift off of an aeroplane is based on this principle.

Chapter 10

Heat

The science of heat and thermodynamics also grows out of the human sense of perception of touch of sight. The subjective notions of heat and cold are replaced by the objective criterion of temperature and gradually there is the realization that heat as a form of energy is manifested as collective mechanical energy of atoms and molecules as well as the energy of radiant or electromagnetic waves. The study of heat and thermodynamics can also be done on similar lines as the principles of mechanics and electromagnetic fields phenomena.

TEMPERATURE AND THERMOMETERS

We are aware of notions of 'hot' and 'cold' through our sense of touch; but this is a subjective and misleading notion. For, if the left hand of a person is kept immersed in cold water and the right hand in hot water for a sufficiently long period of time (few minutes) and then both hands are immersed in luke warm water the resulting sense of perception of the left hand is that the 'luke warm' water is hot and that of right hand is that the same 'luke warm' is cold.

The first step towards an objective measure of temperature is to determine the unique property of some substance like mercury, to which temperature can be uniformly co-related. Mercury, in the form of a thread in a narrow bore of a glass tube expands uniformly over the same interval of change in its temperature. This thermometric property of mercury is used to define three temperature scales, Celsius, Fahrenheit and Roemer. In each of these scales the melting point of ice and the boiling point of water, under 1 atmospheric pressure, are used as standard references.

CELSIUS SCALE

The melting point of ice is taken as 0^0C and the boiling point of water is taken as 100^0C, and the interval between these two reference levels is divided into 100 equal divisions, each division is called 1^0C (1^0 Centigrade or 1^0 Celsius).

FAHRENHEIT SCALE

The melting point of ice is taken as 32^0F and the boiling point of water is taken as 212^0F; the interval between these levels is divided into 180 equal parts, each part is called 1^0F (1^0Fahrenheit).

ROEMER SCALE

The melting point of ice is taken as 0^0R and the boiling point of water is taken as 80^0R; the interval between these levels is divided into 80 equal parts, each part is called 1^0R (1^0 Roemer).

The temperature from one scale to the other can be converted by using the relation,

$$\frac{F-32}{180} = \frac{c}{100}$$

$$= \frac{R}{80}$$

Thus, e.g., $62°\text{F} = \frac{100°c}{6} = \frac{80°R}{6}$

$$10°c = 50°F$$

$$= 8°\text{R}$$

$$5^0 R = \frac{100°C}{80}$$

$$= 32 + \frac{180}{80} \times 5$$

$$= 42.25°F$$

Now a days Celsius scale or Kelvin scale is used in scientific work; Fahrenheit scale is used largely in clinical thermometers; whereas Roemer scale is not used anywhere.

Solved Problem

1. If the temperature of an object is 50^0C, then what will this temperature be in ^0F and ^0R ?

Solution:

$$\frac{C}{100} = \frac{F-32}{180}$$

$$= \frac{R}{80}$$

$$\therefore F = 32 + \frac{9}{5} \times C \; R$$

$$R = 0.8C$$

$$\therefore F = 32 + \frac{9}{5} \times 50$$

$$R = 0.8 \times 50$$

$$\therefore F = 32 + 90$$

$$R = 40$$

$$\therefore F = 122$$

$$R = 40$$

$\therefore$ Centrigrade temp 50^0C is 122^0 F or 40^0 R.

THERMAL EXPANSIONS OF SOLIDS, LIQUIDS AND GASES

It is known from experience that matter in all three states solids, liquids and gases expands when heated. The characteristics of expansion are considered in details as follows.

SOLIDS

Solids are available in definite shapes of rods, laminas and blocks. Therefore the following three coefficients can be defined.

COEFFICIENT OF LINEAR EXPANSION: α

The change in length, per unit original length per 1^0C change in temperature, is called the coefficient of linear expansion.

$$i.e., \alpha = \frac{\Delta l}{1 \times \Delta\theta} / C^0 \text{or} \left(C^0\right)^{-1}$$

Note:

1. Expansions in surface area and volume are negligible.
2. α varies slightly over a range of temperatures, but the variation is negligibly small.

Coefficient of Superficial Expansion: β

The change in surface area per unit original area per 1^0 C change in temperature is called the coefficient of superficial expansions.

$$i.e., \beta = \frac{\Delta A}{A \times \Delta\theta} / C^0 \text{or} \left(C^0\right)^{-1}$$

Note:

1. Expansion in volume is negligible.
2. Same comment as for α
3. $\beta \cong 2\,\alpha$

Coefficient of Cubical Expansion

The change in volume per unit original volume per 1^0C change in temperature is called coefficient of cubical expansion.

$$i.e. \gamma = \frac{\Delta V}{V \times \Delta\theta} / C^0 or \left(C_0\right)^{-1}$$

Note:

1. Same comment for variation in γ? as for α & β.
2. $\gamma \cong 3\alpha$.

LIQUIDS

Liquids have only definite volume but no definite shape, hence only volume change occurs with change in temperature. Therefore, only coefficient of cubical expansion can be defined for liquids.

It is defined as change in volume per unit original volume per 1^0 C change in temperature.

$$i.e. \gamma = \frac{\Delta V}{V \times \Delta\theta} / C^0$$

However, liquids have to be taken in containers. Their observed expansion is lesser than their real expansion, that would have occurred, if the containers were not to expand; therefore γ_r: Real coefficient & γ_a: Apparent coefficients have to be distinguished and their co-relation is given by

$$\gamma_r \cong \gamma_a + \gamma_c$$

where γ_c: Coefficient of cubical expansion of container.

GASES

Gases neither have definite size (volume) nor definite shape. Moreover, the changes in pressure are equally important (in case of solids & liquids these changes are negligible) as the changes in volume, when gases are heated.

Therefore to determine the unique relation between changes in volume and temperature, the pressure must be kept constant and to determine unique relation between changes in pressure and temperature the volume must be kept constant. Hence the following two coefficients are defined for enclosed masses of gases.

Coefficient of Volume Expansion at Constant Pressure: γ_p

The change in volume, above the volume at 0^0 C per unit original volume at 0 ^{0} C per 1^0 C change in volume, at constant pressure, is called γ_p.

Let V_0 = Volume of enclosed mass of gas at 0°c (at constant pressure)

V: Volume of enclosed mass of gas at t°C

$$\text{Then } \gamma_p = \frac{V - V_0}{V_0 \times (t-0)} / C_0$$

$$\text{or } V = V_0 \left[1 + \gamma_p \times t\right]$$

Note: Experimentally it is found that $\gamma_p = \frac{1}{273} / C_0$ for all known gases.

Coefficient of Pressure Expansion at Constant Volume: γ_v

The changes in pressure, above the pressure at 0^0C per unit original pressure at 0^0C per 1^0C change in temperature, at constant volume, is called γ_v.

Let P_0: Pressure of enclosed mass of gas 0°C (at constant volume)

P: Pressure of enclosed mass of gas at t^0C

$$\text{Then } \gamma_v = \frac{P - P_0}{P_0 \times (t-0)} / {}^0C$$

$$\text{or } P = P_0 \left[1 + \gamma_v \times t\right]$$

Note: Experimentally it is found that $\gamma_v = 1/C^0$ for all known gases.

Kelvin or Absolute or Ideal Gas Scale of Temperature

consider $\gamma_p = \frac{1}{273}/C^0$

and $V = V_0[1 + \gamma \times t]$

i.e. $V = V_0\left[1 + \frac{t}{273}\right]$

i.e. $V = V_0\left[\frac{273 + t}{273}\right]$

Therefore at t = –273⁰C the volume of gas becomes V = 0, which must be treated as the limit. Thus, if a gas were to always remain a gas (therefore called ideal gas) without getting liquefied then the lowest possible temperature to which it can be cooled down must be -273^0C and not less than that (for the volume then will become negative; clearly absurd proposal).

Although, the Celsius Scale of temperature is an objective measure, it is still dependent on properties of one particular substance viz. water; whereas all ideals gases tend to attain V = 0 at t = -273^0C. Therefore, it is better to redefine the temperature scale in terms of behaviour of all the ideal gases, irrespective of their chemical nature, rather than in terms of one particular substance viz. water.

The new temperature scale was introduced by Lord Kelvin, hence it is called the Kelvin Scale of temperature and it is defined as

T^0 K = t = 273, where t is the temperature on the Celsius Scale.

Thus, T = –273 + 273 = 0^0K at $(t)_{lowest} = -273^0$C.

It is obvious that just as -273^0C is the lowest possible temperature that can occur in nature, 0^0K is also the lowest possible temperature; hence 0^0K is the absolute zero of temperature. Since Kelvin Scale of temperature is based on the behaviour of Ideal gases, it is also known as the Ideal gas scale of temperatures. A scale of temperature can be defined independent of any substance; this scale of temperature can be called Absolute Scale and it will be shown that it is uniquely related to Kelvin Scale, hence Kelvin Scale is called also as Absolute Scale of temperatures.

GAS LAWS

The behaviour of enclosed masses of ideal gases is determined by the relations between P_1 V or P_1T, or V_1T when the third quantity T or V or P respectively, is maintained constant; these relations were obtained experimentally by Boyle, Gay-Lussac, Charles respectively.

BOYLE'S LAW

The Pressure (P) of an ideal gas of enclosed mass varies inversely with its Volume (V) if temperature (T) is kept constant.

i.e., $P\alpha\frac{1}{V}$ at T= Constant.

i.e., PV = Constant, at T = Constant

or $P_1V_1 = P_2V_2$ at T = Constant

GAY-LUSSAC'S LAW

The pressure (P) of enclosed mass of an ideal gas varies directly with its temperature (T) if volume (V) is kept constant.

i.e., $P\alpha T$, at V = Constant

i.e., $\dfrac{P}{T}$ = Constant at V = Constant

i.e., $\dfrac{P_1}{T_1} = \dfrac{P_2}{T_2}$ at V = Constant

Note: This law is obvious from the definition of γ_v.

For $P = P_0\left[\dfrac{t+273}{273}\right]$(because $\gamma_r = \dfrac{1}{273}/C_0$

Put $t+273 = T^0 K$

$$P = P_0\left[\frac{T}{T_0}\right]$$

or $\dfrac{P_1}{T_1} = \dfrac{P_2}{T_2}$

CHARLES' LAW

The volume (V) of an enclosed mass of ideal gas varies directly with its temperature (T) if pressure (P) is kept constant.

i.e., $V \alpha T$ at P = Constant

i.e., $\dfrac{V}{T}$=Constant, at P = Constant

i.e., $\dfrac{V_1}{T_1} = \dfrac{V_2}{T_2}$ at P = Constant

Note: This law is obvious from definition of γ_p.

For $V = V_0\left[\dfrac{t+273}{273}\right]$ (because $\gamma_p = \dfrac{1}{273}/C_0$)

$$\therefore V = V_0\left[\frac{T}{T_0}\right]$$

$$\therefore \frac{V}{T}\frac{V_0}{T_0} \text{ or } \frac{V_1}{T_1} = \frac{V_2}{T_2}$$

EQUATION OF STATE: IDEAL GAS EQUATION

The behaviour of ideal gases (of enclosed mass) is characterized in terms of an unique relation between $P_1V_1T_1$; Such an equation is called the Equation of State of Ideal gas. Generally the behaviour of any state of matter can be characterized by some unique relation between Pressure (P) and density (ρ) - which of course corresponds to volume (V). The equation of the State of ideal gases is also known as Ideal Gas equation and it can obtained by the combination of any two of the three gas laws as stated earlier.

IDEAL GAS EQUATION

Let an enclosed mass of gas initially be in the state P_1, V_1, T_1 and let it be taken to some final state P_2, V_2, T_2, through some intermediate state, P_1', V_1', T_1' as follows:

$$P_1, V_1, T_1 \rightarrow P_1', V_1', T_1' \text{ under constant pressure}$$

$$\therefore P_1 = P_1' \,\&\, \frac{V_1}{T_1} = \frac{V_1'}{T_1'}, \text{ from Charles' law}$$

$$P_1', V_1', T_1' \rightarrow P_2, V_2, T_2 \text{ under constant volume}$$

$$\therefore V_1' = V_2 \,\&\, \frac{P_1'}{T_1'} = \frac{P_2}{T_2} \text{ from Gay-Lussac's law}$$

$$\text{Therefore, } \frac{V_1}{T_1} = \frac{V_1'}{T_1'} \; i.e. \frac{V_1}{T_1'} = \frac{T_1}{T_1'} \,\&\, V_1' = V_2$$

$$\& \; \frac{V_1'}{T_1'} = \frac{P_2}{T_2} \text{ i.e., } \frac{P_1'}{P_2} = \frac{T_1'}{T_2} \;\&\; V_1' = V_2$$

$$\therefore \frac{V_1}{V_1'} \times \frac{P_1'}{P_2} = \frac{T_1}{T_1'} \times \frac{T_1'}{T_2}$$

$$\therefore \frac{V_1}{V_1'(=V_2)} \times \frac{P_1'(=P_1)}{P_2} = \frac{T_1}{T_2}$$

$$\therefore \frac{V_1 \times P_1}{V_2 \times P_2} = \frac{T_1}{T_2}$$

$$\text{or } \frac{PV}{T} = \text{Constant}$$

Note:

It is founded that the value of constant is dependent on the enclosed mass of gas for a given gas and is also found to differ for different gases of 1 unit of mass. However, it is found that if 1 mol. mass (Mass numerically equivalent in gm to molecular weight, e.g., 2 gm for H_2, 32 gm for O_2, 28 gm for N_2 etc.) of any ideal gas is chosen then the value of the constant is same for all gases. This universal constant (same for all gases) is denoted usually by 'R' and is called the universal gas constant.

The ideal gas equation therefore is normally written as

$PV = nRT$ where n = number of molecules.

The no. of molecules is defined as, the ratio of actual mass of gas M to its molecular weight (M_o)

i.e. $n = \frac{M}{M_0}$

If m is the mass of each molecule of a given gas and N is the no.of molecules that make up the mass = M; whereas N_o= Avogadro's no. = Number of molecules in 1 mole of the gas (any gas), then $M = mN$ & $M_o = mN_o$.

Therefore $\frac{N}{N_0}$

$$\therefore \quad PV = nRT = \frac{M}{M_0}RT = \frac{N}{N_0}RT$$

$$\text{Now, } PV = \frac{M}{M_0}RT$$

$$\therefore \frac{PV}{M} = \frac{R}{M_0}T$$

$$\therefore \frac{P}{\frac{M}{V}} = \frac{R}{mN_0}T$$

$$\text{but } \frac{M}{V} = \rho = \text{density of gas}$$

$$\text{and } \frac{R}{N_0} = K = \text{Boltzman's constant}$$

$$\therefore \frac{P}{\rho} = \frac{KT}{m}$$

$$\text{also } PV = \frac{N}{N_0}RT = N\left(\frac{R}{N_0}\right)T = NKT$$

$$\therefore PV = NKT$$

KINETIC THEORY OF IDEAL GASES

The concept of atoms and molecules being the ultimate building blocks of bulk matter was revived by Dalton in order to explain the three laws of chemical combinations. In the mid 19th century, these concepts, together with certain assumptions regarding the nature, size, distribution and motions of the molecules and atoms, were synthesized with Newtonian mechanics, in order to explain the observed behaviour of ideal gases. This work by Maxwell, Boltzman and others, led to the development of what is known as the Kinetic Theory of gases.

The Assumptions of the Kinetic Theory are

1. Any gas is made up of a very large number of molecules.
2. Molecules of a gas are identical to each other, with respect to shape, size and mass.
3. The molecules are perfectly rigid spheres of negligible radius.
4. The molecules are in an incessant state of chaotic motion in all possible speeds and directions.
5. The distribution of molecules is homogenous and isotropic in any container enclosing the gas.
6. The molecules exert force on each other only when they collide with each other or with the walls of the container.
7. The collision between molecules or molecules and the walls of the container are perfectly elastic collisions, i.e., only such collisions can change the directions of motion but not their speeds.
8. Between successive collisions the molecules travel freely with constant velocities; the distance travelled freely is called free path. On an average, the mean free path of all molecules is the same.
9. The average Kinetic energy of a molecule is proportional to the absolute temperature of the gas.

EXPRESSION FOR PRESSURE EXERTED BY GAS

Let N number of molecules of an ideal gas of mass M, be enclosed in a cubical base of sides L. Let ith molecule move with velocity C_i, and u_i, v_i, w_i be its X, Y, Z components.

Then $c_i^2 = u_i^2 + v_i^2 + w_i^2$

Consider only X-component of motion of ith molecule, then the force exerted by ith molecule, on account of its periodic collisions at $\frac{2L}{u_i}$ with front wall and change in its momentum of -2*mu* $_i$ is

$$f_{ix} = \frac{\frac{+2mu_i}{\frac{2L}{u_i}}}{}$$

$$= \frac{+mu_i^2}{L}$$

∴ Force on front wall by all '*N*' molecules

is $F_x = \sum_{i=1}^{N} f_{ix} = \sum_{i=1}^{N} \frac{mu_i^2}{L}$

∴ Pressure on front wall is

$$P_x = \frac{F_x}{L^2}$$

$$= \frac{\sum_{i=1}^{N} \frac{mu_i^2}{L}}{L^2}$$

i.e., $Px = \left(\frac{m}{L^3}\right)$

$$= \left(\frac{m}{L^3}\right)\left(\sum_{i=1}^{N} u_i^2\right)$$

i.e., $Px = \left(\frac{m}{v}\right)\left(\sum_{i=0}^{N} u_i^2\right)$ (V= L_3= Volume of gas)

Like wise pressure of gas on walls ⊥ *r* to *Y* and *Z* axis are respectively,

$$Py = \left(\frac{m}{v}\right)\left(\sum_{i=0}^{N} v_i^2\right)$$

$$\text{and } Pz = \left(\frac{m}{v}\right)\left(\sum_{i=1}^{N} w_i^2\right)$$

$$\therefore P_x + P_y + P_z = \left(\frac{m}{v}\right)\left(\sum_{i=0}^{N}\left\{u_i^2 + v_i^2 + w_i^2\right\}\right)$$

$$\therefore 3p = \left(\frac{m}{v}\right)\left(\sum_{i=0}^{N} c_i^2\right)$$

For convenience only we have treated $P_{x'}$ $P_{y'}$ P_z are different, or we have determined pressure thrice over as all N molecules were considered to be moving along *X*, *Y* and *Z* all the three axis.

$$\therefore P = \frac{1}{3}\left(\frac{m}{v}\right)\left(\sum_{i=1}^{N} c_i^2\right)$$

Now, let uis define

$$\overline{C^2} = \text{Mean Square velocity of gas molecules}$$

$$\therefore \overline{C^2} = \left(\frac{\sum_{i=1}^{N} c_i^2}{N}\right)$$

$$\left(\sum_{i=0}^{N} c_i^2\right) = N\overline{C^2}$$

$$\therefore P = \frac{1}{3}\left(\frac{m}{V}\right)\left(N\overline{C_2}\right)$$

$$\therefore P = \frac{1}{3}\frac{(mN)}{V}\times\overline{C^2}$$

but $mN = M$ = Mass of gas

$$\therefore P = \frac{1}{3}\frac{(M)}{V}\times\overline{C^2}$$

$$\text{But } \rho = \frac{M}{V} = \text{Density of gas}$$

$$\therefore P = \frac{1}{3}\rho c^2$$

Finally, we define Root Mean Square (R.M.S) velocity C of molecules as Square root of the Mean Square Velocity.

$$\text{i.e, } C = \sqrt{\overline{C^2}}$$

$$C^2 = \overline{C^2}$$

$$\therefore P = \frac{1}{3}\rho C^2$$

It is customary to call this equation, i.e., pressure expression, as the equation of state. In the above derivation of pressure expression, all but the last assumption viz, Average K.E of any molecules is proportional to the absolute temperature of the gas, has been invoked at some stage or the other.

IDEAL GAS EQUATION FROM KINETIC THEORY

$$\text{consider } P = \frac{1}{3}\rho C^2$$

$$= \frac{1}{3}\frac{M}{V}C^2$$

$$= \frac{1}{3}\frac{mN}{V}\overline{C^2}$$

$$= \frac{2}{3}\left(\frac{N}{V}\right)\left(\frac{1}{2}m\overline{C^2}\right)$$

but $\frac{1}{2}m\overline{C^2}$ = Average K.E of a molecule α T

$\therefore \frac{1}{2}m\overline{C^2} = \frac{3}{2}KT$ (the choice of constant as 3/2 K is mandatory to obtain ideal gas equation in the same format as earlier)

$$P = \frac{\cancel{2}}{\cancel{3}}\left(\frac{N}{V}\right)\left(\frac{\cancel{3}}{\cancel{2}}KT\right)$$

$$\therefore PV = NKT$$

$$\therefore \text{PV} = N\left(\frac{R}{N_0}\right)^T$$

$$= \left(\frac{N}{N_0}\right)^{RT}$$

$$\therefore PV = nRT$$

The assumption $\frac{1}{2}m\overline{C^2} = \frac{3}{2}KT$ implies the interpretation of heat energy as mechanical energy of the molecules, albeit as statistical concept only; that is, temperature is the manifestation of the average motion of a large number of molecules; it is absurd to say that $\frac{1}{2}mc_i^2 = \frac{3}{2}KT$ for any i.

Solved Problems

1. A hollow metal sphere is floating on water at 0^0C. If the temperature of water rises to θ^0C, the sphere just submerges in water. Neglect the expansion of sphere. Find expression for determining coefficient of cubical expansion of water.

Solution:

Given d: density of sphere, ρ: density of water, γ: Coefficient of cubical expansion of liquid.

$$(\rho_\theta)\text{water} = (d)\text{sphere}$$

$$\text{but } \rho_\theta = \rho_0[1-\gamma\theta]$$

$$\therefore \rho_0[1-\gamma_\theta] = d$$

$$\therefore \gamma = \frac{\rho_0 - d}{\theta d}$$

2. Estimate the temperature at which oxygen molecules can escape from Earth. (g = 10 m/s, $R_e = 6.4 \times 10^6\ m$, R = 8.3 Joule/mole × ^{0}K, Molecular wt. of O_2 = 32)

Solution:

Given: C = R.M.S. velocity, V_e = escape velocity from Earth then the molecules can escape.

If C = V_e then the molecules can escape.

$$\text{Now, } C = \sqrt{\frac{3RT}{M_0}} \ \& \ V_e = \sqrt{2gR_e}$$

$$\therefore \frac{3RT}{M_0} = 2gR_e$$

$$\therefore T = \frac{2gR_eM_0}{3R}$$

$$T = \frac{2 \times 10 \times 6.4 \times 10^6 \times 32 \times 10^{-3}}{3 \times 8.3}$$

$$T = \frac{(6.4)^2}{3 \times 8.3} \times 10^5$$

$$T = 1.65 \times 10^5 \text{ K}$$

CALORIMETRY AND MECHANICAL EQUIVALENT OF HEAT

The change in internal energy of a system, when heat energy is exchanged with it, and ignore the changes in the external energy of the system. When the substance remains in the same state then the change in internal energy is proportional to the change in temperature, whereas when the substance changes its state (eg. from solid to liquid) the change in internal energy is manifested in terms of the difference of internal energy in the two states and not as the change in temperature.

Heat Capacity (H)

The amount of heat required to change the temperature by 1^0K of a substance is called its Heat Capacity.

i.e $H = \frac{\Delta Q}{\Delta \theta} = \frac{Q}{\theta} Joule / {}^0K$

Q: Amount of Heat energy

θ: temperature

ΔQ: change in the amount of heat energy

$\Delta \theta$: change in the temperature

Specific Heat Capacity (C)

Heat Capacity per unit mass of a substance or the amount of heat required to change the temperature by 1^0K of a unit mass of substance is called Specific Heat Capacity.

$$\text{i.e } C = \frac{H}{m} = \frac{\Delta Q}{m\Delta\theta} Joule / kg^0 K$$

Water Equivalent: (W)

For any given substance the equivalent mass of water W having the same specific heat as the given substance is called its Water Equivalent.

Let H: Heat Capacity of the given substance

W: Mass of water

C_{w}: Specific Heat Capacity of water

$\therefore H_{w}$: Heat Capacity of water

$$H_w = W \times C_w$$

$$\text{but } H_w = H$$

$$\therefore H = W \times C_w$$

$$\therefore W = \frac{H}{C_w}$$

but $H = mC$ m: mass & C: specific heat capacity of a given substance

$$\therefore W = \frac{mC}{C_w}$$

Note: If a unit of specific heat capacity is taken as 1 Calorie / gm × ^{0}C, then C_w = 1 Calorie / *gm* ^{0}C, and therefore

$W = mC$ where W: mass of water in gm.

During changes of state - from solid to liquid and vice versa or from liquid to vapour and vice versa the temperature remains constant but the heat energy is required to be exchanged. The amount of heat energy exchanged per unit mass of a substance is called Latent Heat.

Latent Heat of Fusion or Liquefaction

The amount of heat required per unit mass of a substance at melting point (constant) and under 1 Atmosphere pressure to convert it from the solid to the liquid state is called Latent Heat of Fusion.

$$\text{i.e. } L = \frac{Q}{m} \text{ calorie/gm or K-Calorie/Kg}$$

Latent Heat of Freezing or Solidification

The amount of heat required per unit mass of a substance at freezing point (constant) and under 1 Atmosphere pressure to convert it from the liquid to the solid state is called Latent Heat of Freezing.

$$\text{i.e. } L = \frac{Q}{m} \text{ K-Calorie/Kg}$$

e.g. At 0^0C & 1 Atmosphere pressure 80 *K*-calorie of heat supplied to 1 Kg of ice melts it into water and 80 *K*-Calorie of heat taken away from 1 Kg of water converts it into ice.

$\therefore L = 80$ Kcal / Kg = Latent heat of fusion of ice

Latent heat of freezing of water

Latent Heat of Vapourization

The amount of heat required to vapourize a unit mass of liquid at its boiling point (constant) and under 1 Atmosphere pressure it is called Latent Heat of Vapourization.

$$\text{i.e. } L = \frac{Q}{m} \text{ K-Cal/Kg}$$

Latent Heat of Condensation

The amount of heat required to convert a unit mass of gas from its gaseous state into its liquid state at its condensation point (constant) and under 1 Atmosphere pressure is called Latent Heat of Condensation.

$$\text{i.e. } L = \frac{Q}{m} \text{ K-Cal/Kg}$$

e.g. 540 *K*-Cal of heat supplied under 1 Atmosphere to 1 Kg of water at 100^0C converts it into steam conversely 540 K-Cal of heat taken away from steam at 100^0C and 1 Atmosphere converts it into water.

$\therefore$ L = Latent Heat of vapourization or condensation of water or steam

= 540 K-Cal / Kg.

MECHANICAL EQUIVALENT OF HEAT

Historical development of Mechanism and Heat lead to different units of measurements for mechanical energy and heat energy. In the last century it was recognized that heat is also a form of energy and it can be converted into mechanical energy and vice versa. From Joule's experiments the correlation between the units of 1 Joule and 1 Calorie was found to be

1 Calorie = 4.18 Joule

Therefore, the numerical relation between the amount of work or mechanical energy W (Joule) and amount of heat energy Q (Calorie) can be expressed as $W = JQ$ where J = 4.18 Joule / Calorie. The constant J is called Mechanical Equivalent of heat. It is also known as Joule's constant.

Solved Problem

1. How much heat is released when steam of mass 1 gm at 100^0 C is converted to ice at 0^0C? 'L' of condensation of steam is 540 calorie / gm, freezing of water is 80 calorie / gm and specific heat of water is 1 calorie / gm $\times$ ^{0}C.

Solution:

Q = Heat Released by steam at 100^0 C to become ice at 0^0C.

= Heat Released from steam to water at 100^0C + Heat Released from water to cool from 100^0C to 0^0C + Heat Released from water to freeze into ice at 0^0C.

$$= m \times L_{vapourisation} + m \times c \times (100 - 0) + m \times L_{fusion}$$
$$= 1 \times 540 + 1 \times 1 \times 100 + 1 \times 80$$
$$= 720 \text{ calories}$$

HEAT TRANSFER

From observations and experiments it is found that heat energy can be transferred from one position to another through three different modes: conduction, convection and radiation.

Conduction

This process occurs most significantly in solids. The atoms or molecules in a solid do not leave their mean positions, but continue to vibrate about their mean positions. They transfer heat energy from one atom to another. This happens because of the coupling between them due to mutually attractive forces.

When a solid is heated for a long interval it remains in an unsteady state of heat conduction. The temperature at each point along its length keeps on increasing with time, whereas the temperature along its length decreases from the hot to the cold end, i.e. decreasing towards the end that is away from the source.

After the initial unsteady state, the solid reaches a steady state of heat conduction, therefore variation in temperature with time does not occur. However, with regard to its length, the same condition is maintained i.e. the temperature continues to decrease with increase in its length, as one moves away from the source of heat.

At our level one can understand the process to be a partial absorption of heat energy during the unsteady state by each section and a partial transfer to the next. On the other hand, in a steady state there is no absorption and a complete transfer of heat energy.

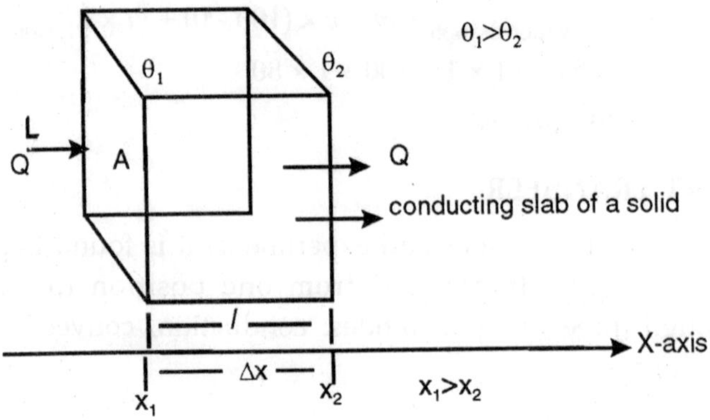

A: Area of Cross- section

l: Δ x = thickness

v_1: Temperature of the left face of the solid where the heat approaches.

v_2: Temperature of the right face from where heat leaves.

Q: Amount of heat

t: time for which the heat is conducted through the slab.

Then in the steady state,

$$Q = KA\left(\frac{\theta_1 - \theta_2}{l}\right)t$$

$$= KA\left(-\frac{\Delta\theta}{\Delta x}\right)t$$

Note: $\Delta\theta = \theta_2 - \theta_1 < 0 \,\&\, \Delta x = l = x_2 - x_1 > 0$

The constant 'K' in the above expression is called the coefficient of thermal conductivity. Its unit is 1 k-cal / m sec ^{0}K or 1 watt / m ^{0}K.

Convection

The process of heat transfer most significant in fluids (liquids and gases) is called convection. In this process the atoms or molecules migrate from one to place another. Thus transfer of heat occurs due to transfer of heated matter. It is readily observed in the form of convection currents in the liquid. One can think of coloured water moving from regions of temperature to lower ones.

Radiation

The process of heat transfer which occurs through empty space and can also occur in matter, in the form of electromagnetic waves is called radiation or radiant heat.

When Q amount of radiation occurs on any object, the object partly reflects (Q_r), partly transmits Q_t) and absorbs the remaining part (Q_a) of heat.

It is obvious that

$$Q_a + Q_r + Q_t = Q(\text{Energy Conservation})$$

$$\therefore \frac{Q_a}{Q} + \frac{Q_r}{Q} + \frac{Q_t}{Q} = 1$$

$$\text{or } a + rt + t = 1$$

The quantities $a = \frac{Q_a}{Q}$, $r = \frac{Q_r}{Q}$, $t = \frac{Q_t}{Q}$ are called respectively coefficients of absorption, reflection and transmission, of radiant heat.

PERFECTLY BLACK BODY

An object whose a = 1, and therefore r = t = 0, i.e. $Q_a = Q$ or which absorbs all the incident radiant heat on it is called a Perfectly black body. Lamp black is closest to a perfectly black body (a = 0.98), but no object in our environment is otherwise found to be a perfectly black body. An artificially constructed perfectly black body is an important source of radiation. The analysis of the spectrum of such an artificial black body leads to Planck's quantum hypothesis.

KIRCHOFF'S LAW OF HEAT RADIATION

This law states that an object is as good an emitter of heat radiation as an absorber of it. In other words, the coefficient of absorption and emissivity of any object are equal to each other; i.e. a = e. This law can be verified by using Rritchie's apparatus.

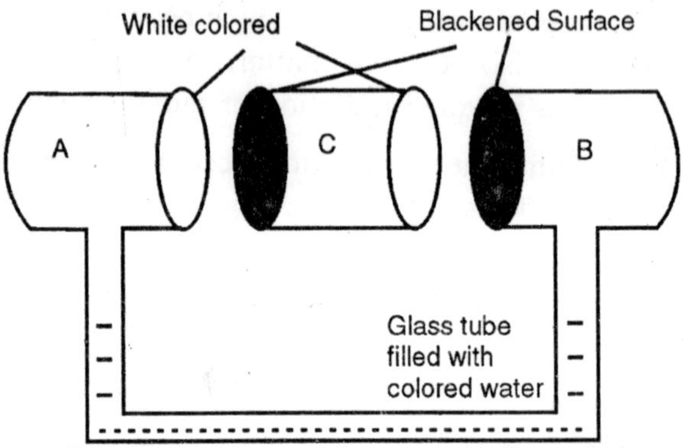

A,B,C : identical & co-axially mounted hollow cylinders

The water in the arms of the glass tube is brought to an equal level before the experiment begins. Then some hot water is poured into the cylinder C. After a sufficiently long interval of time, it is observed that the water levels remain at their original level. The following observation is made:

The fraction of heat a bsorbed by the white surface of A from the blackened surface of C	The fraction of heat absorbed by the blackened surface of B form the white surface of C (in the same time).
$\therefore a$	= e

EMISSIVE POWER: E

The rate of radiation per unit surface area of an object is called its emissive power.

i.e. $E = \dfrac{Q}{AT}$ Q: Amount of heat radiated (of all wave length

A: Surface area

t : time in which heat is radiated

Unit of E-1k-Cal/m^2×sec or 1 watt/m^2

EMISSIVITY OR COEFFICIENT OF EMISSION: e

The ratio of emissive power of any object (E) to the corresponding emissive power of a perfectly black body (E_b) at same temperature is called emissivity.

$$\text{i.e. } e = \frac{E}{E_b}$$

STEFAN'S LAW OF HEAT RADIATION

The emissive power of a perfectly black body is proportional to the fourth power of its absolute temperature.

i.e. $E = \sigma T^4$

where σ = Stefan's constant = 5.67×10^{-8} watt / $m^2 \times K^4$

For any other object $E = e\,\sigma\,T^4$

NEWTON'S LAW OF COOLING

The rate of cooling of an object is directly proportional to the excess temperature of the object over that of its surrounding objects; provided that the excess temperature is sufficiently small.

$$\text{i.e. } \frac{d\theta}{dt} = -K(\theta - \theta_0)$$

K > θ & constant for the object under a given temperature of the surroundings

θ: temperature of the object is ^{0}C

θ_0: temperature of the surrounding objects in ^{0}C

(The surrounding object is assumed to be perfectly black bodies)

If $T << 2T_0$ where T and T_0 are temperatures of the object and the surrounding objects in ^{0}K respectively, then from Stefan's law we can show that

$$\frac{d\theta}{dt} = -\left(\frac{4e\sigma A T_0^3}{mc}\right)(\theta - \theta_0)$$

Chapter 11

Thermodynamics

The changes in a system with specific reference to interaction with the surroundings through exchange of heat energy constitutes study of thermodynamics for the present purpose.

These interactions obey certain laws, which can be distilled from observations and experiments. The science of thermodynamics is largely phenomenological and as such hardly pays any attention to details of constitution of a thermodynamical system. Attempts to derive laws of thermodynamics as consequences of Newton's laws of motion have never been successful though statistical mechanics based on Newtonian laws incorporates a substantial part of thermodynamics.

The scope of this book permits study of thermodynamics essentially in terms of empirically defined and measured quantities like Heat energy (Q), Temperature (T), Pressure (P) and Work (W).

ZEROTH LAW

The concept of temperature is made objective by this law and is defined as the state of thermodynamical system which determines the direction of the flow of heat.

If heat energy is not exchanged between two systems then they are said to be in thermal equilibrium.

"If a system A is in thermal equilibrium with a system B, then they are at the same temperature; if another system C is also in thermal equilibrium with the system A then B and C are also in thermal equilibrium with each other and the temperature of each is same as that of the other."

The above statement is the Zeroth Law of thermodynamics.

THE FIRST LAW

Every thermodynamic system possesses an interval energy U and has the capacity to do external work W which may be called the external energy. When heat energy is supplied (dQ) to a system then from the law of conservation of energy it follows that the sum total of changes in the internal and external energy of the system must equal the heat energy supplied to the system.

i.e. $dQ = dU + dW$

Equation: The mathematical expression for the First law of Thermodynamics.

Note:

1. dU is an exact differential which just depends on the initial and the final states of the system and not on the path connecting these states.
2. dQ and dW are in exact differentials as they are path dependent between the given initial and final states.

WORK DONE AND SOME APPLICATIONS OF THE FIRST LAW OF THERMODYNAMICS

1. The work done by a gas is infinitesimal change of volume dV against the constant external pressure.

It is given by

$$dW = P\,dV$$

When a system is taken from a state P_1, V_1 to the state P_2, V_2 then total work done by the system is,

$$W = \int_1^2 P \, .dV$$

The work done can also be calculated as area under the P-V diagram

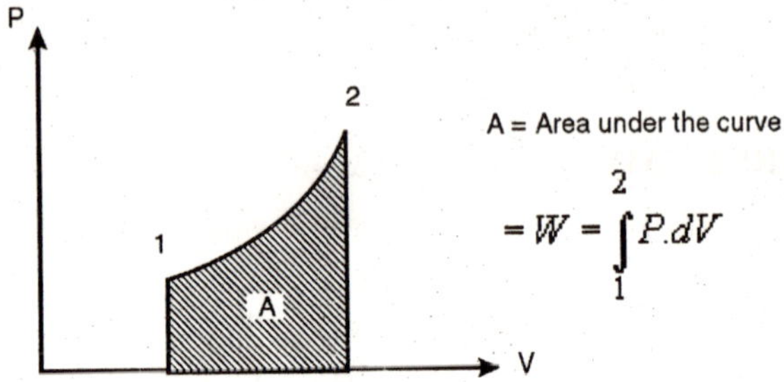

It is obvious, that work done is path dependent; for if a different path is followed in changing from state1 to state2 then the area under the curve will be different and hence work done will be different.

EXPRESSIONS FOR WORK DONE BY AN IDEAL GAS

Let n moles of an ideal gas be made to undergo changes under (a) isochoric (b) isobaric (c) isothermal (d) Adiabatic or isentropic changes.

Then the expressions for work done are as follows:

Isochoric Change:

Constant volume or dV = 0

$$\therefore W = \int_1^2 PdV = 0 \quad \text{as } dV = 0$$

$\therefore W = 0$, when $dV = 0$

Isobaric Change:

Constant Pressure or $dP = 0$

$$\therefore W = \int_1^2 PdV$$

$$= \int_1^2 dV$$

$$= P(V_2 - V_1)$$

$$= nR\,(T_2 - T_1)$$

$$\therefore W = P(V_2 - V_1)$$

$$= nR\,(T_2 - T_1) \quad \text{when } dP = 0$$

Isothermal Change:

Constant Temperature or $dT = 0$

$$\therefore W = \int_1^2 PdV$$

$$= P\int_1^2 \frac{nRT}{V} dV \quad (\therefore\ PV = nRT)$$

$$\left[\therefore p = \frac{nRT}{V}\right]$$

$$\therefore W = nRT\int_1^2 \frac{dV}{V} \quad (\because\ T \text{ is constant})$$

$$\therefore W = nRT\,\text{In}\left(\frac{V_2}{V_1}\right)$$

or using $PV = nRT$, i.e. $V = \dfrac{nRT}{P}$

$$\therefore \frac{V_2}{V_1} = \frac{\dfrac{nRT}{P_2}}{\dfrac{nRT}{P_1}}$$

$$= \frac{P_1}{P_2} \quad (\because\ T = \text{constant})$$

$$W = nRT\,\text{In}\left(\frac{V_2}{V_1}\right)$$

$$= nRT\,\text{in}\left(\frac{P_1}{P_2}\right) \qquad \text{when } d\,T = 0$$

Adiabatic or Isentropic Change

Constant Heat Energy / Entropy

$$\therefore W = \int_1^2 P dV$$

$$= \int_1^2 \frac{K}{V\gamma} dV \quad (\text{since } PV^{\gamma} = K \text{ constant}$$

when dq=0)

$$\therefore W = \frac{K}{-\gamma+1}\left[\frac{V_2}{V_2^{\gamma}} - \frac{V_1}{V_1^{\gamma}}\right]$$

$$\therefore W = \left(\frac{1}{\gamma-1}\right)\left[\left(\frac{K}{V_1^{\gamma}}\right)V_1 - \left(\frac{K}{V_2^{\gamma}}\right)^{V_2}\right]$$

$$\therefore W = \left(\frac{1}{\gamma-1}\right)[P_1V_1 - P_2V_2]$$

$$\therefore W = \left(\frac{1}{\gamma-1}\right)[nRT_2 - nRT_1]$$

$$\therefore W = \frac{nR(T_2 - T_1)}{\gamma-1}, \text{ when dQ} = {}^{0}\!/_{ds} = 0$$

The Adiabatic P-V relation for an ideal gas, dQ = 0

$$dQ = dV + dW$$
$$= dV + P\,dV$$
$$= nC_V\,dT + p\,dV$$
$$= nC_v\left\{\frac{P\,dV + V\,dP}{nR}\right\} + P\,dV \quad \text{Using } PV = nRT$$
$$= \left(\frac{c_v}{R} + 1\right)P\,dV + \left(\frac{C_v}{R}\right)V\,dP$$
$$\therefore dQ = \frac{(C_v + R)P\,dV + (C_v)V\,dP}{R}$$
$$R(dQ) = (C_p)\,P\,dV + (C_v)\,V\,dP$$
$$= (C_v)\left[\left(\frac{C_p}{C_v}\right)P\,dV + V\,dP\right]$$
$$= (C_v)[\gamma P\,dV + V\,dP] \quad \therefore \left(\gamma = \frac{C_p}{C_v}\right)$$

$$\text{but } (dQ) = 0$$
$$\therefore 0 = \gamma P\,dV + V\,dP$$
$$\therefore -\gamma P\,dV = V\,dP$$
$$\therefore -\gamma\frac{dV}{V} = \frac{dP}{P}$$
$$\therefore 0 = \gamma\frac{dV}{V} + \frac{dP}{P}$$
$$\therefore \text{In } K = \gamma \text{ In } V + \text{In } P$$
$$\therefore \text{In } K = \text{In } V^{\gamma} + \text{In } P$$
$$\therefore \text{In } K = \text{In}(PV^{\gamma})$$
$$\therefore PV^{\gamma} = K$$

MEYER'S RELATION

It is well known that ideal gases possess two types of specific heat; specific heat at constant pressure for 1 mole mass C_p and specific heat at constant volume for 1 mole mass. Therefore if 1 mole of ideal gas has its temperature raised by dT^0K, once under isobaric condition and then under isochoric condition then,

$$\therefore C_p dT = (dQ)_p$$

$$= (dV)_p + R\,dT \quad (\because PV=RT \text{ for a mole } \& \; PdV = RdT)$$

$$\& C_v dT = (dP)_v = (dV)_v = 0 \quad (\because dV = 0)$$

$$\because C_p dT - C_v dT = (dQ)_p - (dQ)_v$$

$$= (dV)_p + R\,dT - (dV)_v$$

But $(dV)_p = (dV)_v$ ($\therefore$ dV is the exact differential and deoebds ib dT only)

$$\therefore (C_p - C_v)\,dT = R\,dT$$

$$\therefore C_p - C_v = R$$

Note: R is taken is calorie/mole$\times^0K$ or kilocalorie/-mole$\times^0K$

$$\therefore C_v = \frac{R}{\gamma - 1}$$

$$\therefore C_p = \frac{\gamma^R}{\gamma - 1}$$

THERMODYNAMIC PROCESSES

Reversible Process

If a system is taken from one state to another and can be brought back to the initial state then the process is called reversibility. The change from initial to final state should obviously proceed through the infinitesimal changes, each being a quasi-static state of equilibrium.

Some examples are changes in a gas through isochoric, isothermal or adiabatic changes.

Irreversible Process

If a system is taken from one state to another but cannot be brought back to the same initial state then the process is called irreversible process.

Some examples are free expansion of a gas; dissipation of energy due to friction or the mixing of two gases or liquids etc.

CARNOT CYCLE

Since heat is a form of energy, its conversion to other forms of energy should be possible. The discovery that heat can be converted into mechanical energy was one of the most important cases that let to the Industrial Revolution in the 19th century.

The first law of Thermodynamics allows for the conversion of heat energy into mechanical energy in the same amount. However in order to be able to convert heat energy into mechanical energy in a continuous manner it is necessary to restore the working substance in a device called the heat engine. These device converts heat energy into mechanical energy- back to its original state by taking it through a reversible cyclic process. This cyclic process can operate by transferring some amount of the heat energy absorbed. Thus only a part of the heat energy absorbed is found to be convertible into mechanical energy.

If some enclosed gas is made to absorb heat isothermally, then it will expand and the work done by the gas can be used as mechanical energy but unless the gas is brought back to its starting condition it cannot continue to convert heat energy into mechanical energy. Now, if it is brought back to its original state by compressing it along the reverse path then obviously the mechanical energy obtained during expansion will have to be spent in converting into heat energy; thus the net amount of mechanical energy obtained will be nil.

Therefore, in any heat engine the working substance operates between a source of heat at a higher temperature and a sink of heat at lower energy. Only the difference of heat energy absorbed from a source (or reservoir) and the heat energy transferred to the sink can be converted into mechanical energy. This is true even if the heat engine, i.e. an ideal engine that is no dissipation of energy is assumed to occur. Such an ideal engine is called Carnot's engine and it operates in a Carnot reversible cycle. The working substance used in this is an ideal gas.

The Carnot cycle operates in this manner. The ideal gas of a fixed mass is enclosed in a cylindrical vessel by means of a lightly fitting piston. The mass of the vessel and piston are frictionless. The walls are nonconducting; the base is conducting over which a nonconducting cap can be mounted whenever required. Also a reservoir or a source of heat at temperature $T_1{}^0$K of an unlimited capacity and a sink of heat at temperature $T_2{}^0$K ($T_1 > T_2$) (also of unlimited capacity) are available.

1. Assume the gas to be in the initial state P_1, V_1 at temperature T_1. Let it be placed in thermal contact with the reservoir and allowed to expand isothermally to state P_2, V_2 - absorbing heat energy of amount Q_1 from the reservoir.

2. Take the gas away from reservoir, mount a nonconducting cap on its base, and let it continue to expand adiabatically to the state P_2, V_2 at T^0 K.

3. Now, remove the nonconducting cap and put the gas in thermal contact with sink at $T_2{}^0$ K and compress it isothermally to the state P_3, V_3.

4. Finally, disengage the gas from the sink and mount the non- conducting cap on the base of the cylinder and compress the gas adiabatically to the state P_1, V_1 and $T_1{}^0$ K.

Thus the cycle is complete and gas comes back to the initial state P_1, V_1 at $T_1{}^0$ K. Periodically the cycle of operations (i) and (iv) is repeated to continuously keep on obtain mechanical energy from heat energy.

The cycle is represented on the P- V (indicator) diagram.

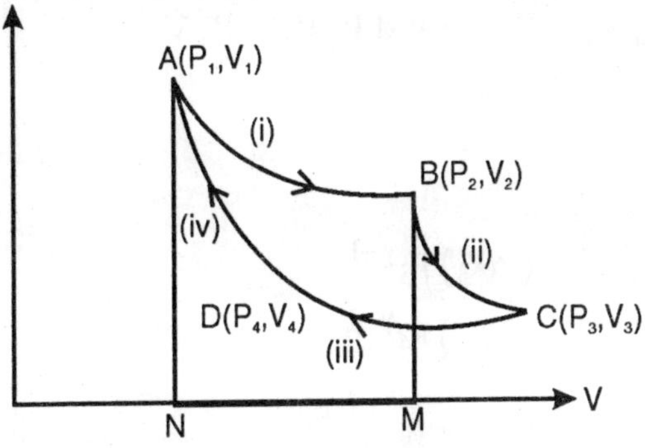

Then

Q_1= Heat absorbed from reservoir during

(i) = $W_1 = nRT_1 \ln\left(\frac{V_2}{V_1}\right)$

Q_2= Heat absorbed during

(ii) = $0 \text{ and } W_2 = \frac{nR}{\gamma - 1}(T_1 - T_2)$

Q_3= Heat transferred to thesink during

(iii) = $W_3 = nRT_1 \ln\left(\frac{V_3}{V_4}\right)$

and Q_4 = Heat transferred to the sink

(iv) = 0, and $W = \frac{nR}{\gamma - 1}(T_1 - T_2)$

$\therefore$ Net heat absorbed by the gas $Q=Q_1+Q_2-Q_3-Q_4$

$$\therefore Q = nRT_1 \text{ in}\left(\frac{V_2}{V_1}\right)+0-nRT_2 \text{ in}\left(\frac{V_3}{V_4}\right)-0$$

$$\therefore Q = nR\left[T_1 \text{ in}\left(\frac{V_2}{V_1}\right)-T_2 \text{ in}\left(\frac{V_3}{V_4}\right)\right]$$

but, $P_2 V_2^\gamma = P_3 V_3^\gamma$ and $P_1 V_1^\gamma = P_2 V_2^\gamma$
and $P_4 V_4^\gamma = P_1 V_1^\gamma$ and $P_4 V_4^\gamma = P_3 V_3^\gamma$

$$\therefore \frac{P_2V_2^\gamma}{P_1V_1^\gamma} = \frac{P_3V_3^\gamma}{P_4V_4^\gamma}$$

$$\therefore \frac{(P_2V_2)V_2^{\gamma-1}}{(P_1V_1^\gamma)V_1^{\gamma-1}} = \frac{(P_3V_3)V_3^{\gamma-1}}{(P_4V_4)V_4^{\gamma-1}}$$

$$\therefore \left(\frac{V_2}{V_1}\right)^{\gamma-1} = \left(\frac{V_3}{V_4}\right)^{\gamma-1}$$

$$\therefore \frac{V_2}{V_1} = \frac{V_3}{V_4}=x$$

$$\therefore Q = nR\ (T_1\text{-}T_1) \text{ in } X$$

And, net work done by the gas = $W= W_1+W_2+W_3+W_4$

$$\therefore W = nRT_1 \text{ in}\left(\frac{V_2}{V_1}\right)+\frac{nR}{\gamma-1}(T_1-T_2)-nRT_2 \text{ in}\left(\frac{V_3}{V_4}\right)$$

$$+\frac{nR}{\gamma-1}(T_2-T_1)$$

$$\therefore W = nR(T_1-T_2)/nX$$

$$\therefore Q = W$$

Now, the efficiency (η) of the Carnot's heat engine can be defined as the ratio of useful mechanical energy (W) obtained to total heat absorbed by the gas (Q_1),

$$\text{i.e. } \eta = \frac{W}{Q_1} = \frac{Q}{Q_1}$$

$$\text{or } \eta = \frac{nR(T_1 - T_2)/nX}{nRT_1/nX}$$

$$\text{or } \eta = \frac{T_1 - T_2}{T_1}$$

$$= 1 - \frac{T_2}{T_1}$$

$$\eta = 1 - \frac{T_2}{T_1}$$

Note: (1) As $T_2 > T_1$ always $\therefore \eta < 1$

(2) Efficiency can also be expressed.

$$\eta = \frac{W}{Q_1}$$

$$= \frac{\text{Area ABCDA}}{\text{Area ABMNA}}$$

THE SECOND LAW

The nature of heat energy as found in the Carnot Cycle operation is such that it cannot be completely converted into mechanical energy since it is a cyclic operations. This observation is universally true, i.e. independent of the design of the engine and the nature of the working substance.

This observation therefore becomes a law of nature called The Second Law of Thermodynamics. It was formulated by Kelvin and Claussius in different but similar statements. According to Kelvin: "No heat engine can be designed which can convert completely all heat energy from a reservoir into mechanical energy without producing any changes in the surroundings." According to Claussius: "It is impossible to cool any object below the coldest object in the surroundings."

The Second Law of Thermodynamics has been formulated in various forms, which are equivalent to each other.

Kelvin's statement

It is impossible to cool an object below the coldest object in its surroundings, without producing any changes in the surroundings.

Claussius's Statement

It is impossible to transfer heat from a lower to a higher temperature without producing any changes in the surroundings.

Planck-Kelvin Statement

No heat engine can be designed which can convert heat from a single reservoir completely into mechanical energy.

Entropy

A quantity called Entropy can be derived from the study of the Carnot Cycle. This quantity can then be used to formulate mathematically The Second Law of Thermodynamics.

When heat Q is exchanged at temperature T then the entropy changes by $ds = \frac{dQ}{T}$ amount

Now, in Carnot Cycle

(1) $(\Delta S)_{(i)}$ = Change in Entropy during isothermal expansion

$$= \int_1^2 ds = \int_1^2 \frac{dQ}{T_1} = \frac{1}{T_1}\int_1^2 dQ = \left(\frac{1}{T_1}\right) \times Q_1$$

$$\therefore (\Delta S)_{(i)} = \frac{nR\cancel{T_1}\, ln\left(\frac{V_2}{V_1}\right)}{\cancel{T_1}}$$

$$= nR\, ln\left(\frac{V_2}{V_1}\right)$$

$$\therefore (\Delta S)_{(i)} = nR\, ln\left(\frac{V_2}{V_1}\right)$$

(2) $\therefore (\Delta S)_{(i)}$ = Change Entropy during adiabatic expansion.

$$= \int_2^3 ds = \int_2^3 \frac{dQ}{T_1} = 0 \qquad \text{(as } dQ = 0\text{)}$$

$$\therefore (\Delta S)_{(ii)} = 0$$

(3) $(\Delta S)_{(iii)}$ = Change in Entropy during adiabatic expansion

$$= \int_3^4 ds = \int_3^4 \frac{dQ}{T} = \frac{1}{T_2}\int_3^4 dQ$$

$$= \left(\frac{1}{\cancel{T_2}}\right) \times nR\cancel{T_2}\, ln\left(\frac{V_4}{V_3}\right)$$

$$\therefore (\Delta S)_{(iii)} = nR\, ln\left(\frac{V_4}{V3}\right)$$

$$= -nR\, ln\left(\frac{V_3}{V_4}\right)$$

$\therefore (\Delta S)_{(iv)}$ = Change in Entropy during adiabatic compression

$$= \int_4^1 ds = \int_4^1 \frac{dQ}{T} = 0 \quad (\text{as } dQ = 0)$$

$$\therefore (\Delta S)_{(iv)} = 0$$

$\therefore$ Net change in Entropy per cycle is

$$\Delta S = (\Delta S)_{(i)} + (\Delta S)_{(ii)} + (\Delta S)_{(iii)} + (\Delta S)_{(iv)}$$

$$= nR/n\left(\frac{V_2}{V_1}\right) + 0 - nR/n\left(\frac{V_3}{V_4}\right) + 0$$

$$= nR\left[\text{in}\left(\frac{V_2}{V_1}\right) - \text{in}\left(\frac{V_3}{V_4}\right)\right]$$

$$\text{but } \frac{V_2}{V_1} = \frac{V_3}{V_4}$$

$$\therefore \Delta S = 0$$

Thus in a reversible thermodynamic process the change in entropy of the system is zero, i.e. the entropy remains constant whereas in an irreversible process we find that change in entropy is always greater than zero.

$$\text{Thus, } \therefore \Delta S \begin{cases} = 0 \text{ Reversible process} \\ > 0 \text{ Irreversible process} \end{cases}$$

Hence, the second Law of Thermodynamics states that entropy of the universe never decreases.

Chapter 12

Radioactivity

STABILITY OF NUCLEI

As mass numbers become larger, the ratio of neutrons to protons in the nucleus becomes larger for the stable nuclei. Non-stable nuclei may have an excess or deficiency of neutrons and undergo a transformation process known as beta (β) decay. Non-stable nuclei can also undergo a variety of other processes such as alpha (α) or neutron (n) decay. As a result of these decay processes, the final nucleus is in a more stable or more tightly bound configuration.

In radioactive processes, particles or electromagnetic radiation are emitted from the nucleus. The most common forms of radiation emitted have been traditionally classified as alpha (α), beta (β), and gamma (γ) radiation. Nuclear radiation occurs in other forms, including the emission of protons or neutrons or spontaneous fission of a massive nucleus.

Of the nuclei found on Earth, the vast majority are stable. This is so because almost all short-lived radioactive nuclei have decayed during the history of the Earth. There are approximately 270 stable isotopes and 50 naturally occurring radioisotopes (radioactive isotopes). Thousands of other radioisotopes have been made in the laboratory.

Radioactive decay will change one nucleus to another if the product nucleus has a greater nuclear binding energy than the initial decaying nucleus. The difference in binding energy (comparing the before and after states) determines which decays are energetically possible and which are not. The excess binding energy appears as kinetic energy or rest mass energy of the decay products.

The Chart of the Nuclides, part of which is shown above is a plot of nuclei as a function of proton number, Z, and neutron number, N. All stable nuclei and known radioactive nuclei, both naturally occurring and manmade, are shown on this chart, along with their decay properties. Nuclei with an excess of protons or neutrons in comparison with the stable nuclei will decay toward the stable nuclei by changing protons into neutrons or neutrons into protons, or else by shedding neutrons or protons either singly or in combination. Nuclei are also unstable if they are excited, that is, not in their lowest energy states. In this case the nucleus can decay by getting rid of its excess energy without changing Z or N by emitting a gamma ray.

Nuclear decay processes must satisfy several conservation laws, meaning that the value of the conserved quantity after the decay, taking into account all the decay products, must equal the same quantity evaluated for the nucleus before the decay. Conserved quantities include total energy (including mass), electric charge, linear and angular momentum, number of nucleons, and lepton number (sum of the number of electrons, neutrinos, positrons and antineutrinos–with antiparticles counting as –1).

137Ba decay data, counting numbers of decays observed in 30-second intervals. The best-fit exponential curve is shown. The points do not fall exactly because of statistical counting fluctuations.

The probability that a particular nucleus will undergo radioactive decay during a fixed length of time does not depend on the age of the nucleus or how it was created. Although the exact lifetime of one particular nucleus cannot be predicted, the mean (or average) lifetime of a sample containing many nuclei of the same isotope can be predicted and measured. A convenient way of determining the lifetime of an isotope is to measure how long it takes for one-half of the nuclei in a sample to decay–this quantity is called the half-life, $t_{1/2}$. Of the original nuclei that did not decay, half will decay if we wait another half-life, leaving one-quarter of the original sample after a total time of two half-lives. After three half-lives, one-eighth of the original sample will remain and so on. Measured half-lives vary from tiny fractions of seconds to billions of years, depending on the isotope.

The number of nuclei in a sample that will decay in a given interval of time is proportional to the number of nuclei in the sample. This condition leads to radioactive decay showing itself as an exponential process. The number, N, of the original nuclei remaining after a time t from an original sample of N_0 nuclei is

$$N = N_0 e^{-(t/T)}$$

where T is the mean lifetime of the parent nuclei. From this relation, it can be shown that $t_{1/2} = 0.693T$.

Nuclear Decay

Whenever a nucleus can attain a more stable (i.e., more tightly bound) configuration by emitting radiation, a spontaneous disintegration process known as radioactive decay or nuclear decay may occur. In practice, this "radiation" may be electromagnetic radiation, particles, or both. Detailed studies of radioactive decay and nuclear reaction processes have led to the formulation of useful conservation principles. The four principles of most interest in this module are:

Conservation of electric charge implies that charges are neither created nor destroyed. Single positive and negative charges may, however, neutralize each other. It is also possible for a neutral particle to produce one charge of each sig

Conservation of mass number does not allow a net change in the number of nucleons. However, the conversion of a proton to a neutron and vice versa is allowed.

Conservation of mass and energy implies that the total of the kinetic energy and the energy equivalent of the mass in a system must be conserved in all decays and reactions. Mass can be converted to energy and energy can be converted to mass, but the sum of mass and energy must be constant.

Conservation of momentum is responsible for the distribution of the available kinetic energy among product nuclei, particles, and/or radiation. The total amount is the same before and after the reaction even though it may be distributed differently among entirely different nuclides and/or particles.

Alpha Decay

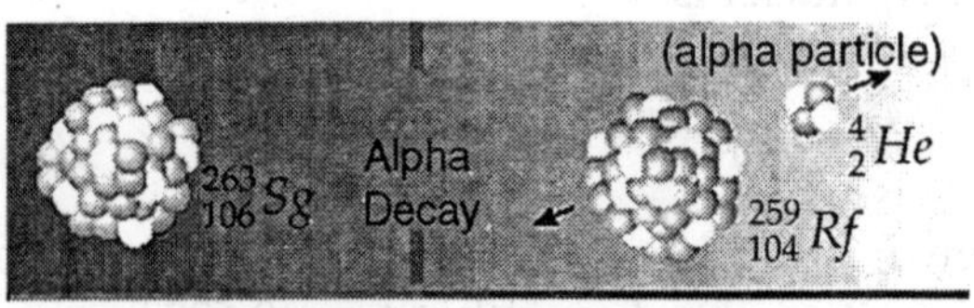

Fig. An alpha-particle decay

In alpha decay, the nucleus emits a 4He nucleus, an *alpha particle*. Alpha decay occurs most often in massive nuclei that have too large a proton to neutron ratio. An alpha particle, with its two protons and two neutrons, is a very stable configuration of particles. Alpha radiation reduces the ratio of protons to neutrons in the parent nucleus, bringing it to a more stable configuration. Many nuclei more massive than lead decay by this method.

Consider the example of ^{210}Po decaying by the emission of an alpha particle. The reaction can be written $_{210}$Po Æ $_{206}$Pb + $_{4}$He. This polonium nucleus has 84 protons and 126 neutrons. The ratio of protons to neutrons is Z/N = 84/126, or 0.667. A $_{206}$Pb nucleus has 82 protons and 124 neutrons, which gives a ratio of 82/124, or 0.661. This small change in the Z/N ratio is enough to put the nucleus into a more stable state, brings the "daughter" nucleus (decay product) into the region of stable nuclei in the Chart of the Nuclides.

In alpha decay, the atomic number changes, so the original (or parent) atoms and the decay-product (or daughter) atoms are different elements and therefore have different chemical properties.

In the alpha decay of a nucleus, the change in binding energy appears as the kinetic energy of the alpha particle and the daughter nucleus. Because this energy must be shared between these two particles, and because the alpha particle and daughter nucleus must have equal and opposite momenta, the emitted alpha particle and recoiling nucleus will each have a well-defined energy after the decay. Because of its smaller mass, most of the kinetic energy goes to the alpha particle.

Beta Decay

Beta particles are electrons or positrons (electrons with positive electric charge, or antielectrons). Beta decay occurs when, in a nucleus with too many protons or too many neutrons, one of the protons or neutrons is transformed into the other. In beta minus decay, a neutron decays into a proton, an electron, and an antineutrino: n → p + e – +. In beta plus decay, a proton decays into a neutron, a positron, and a neutrino: p → n + e+ +*n*. Both reactions occur because in different regions of the Chart of the Nuclides, one or the other will move the product closer to the region of stability.

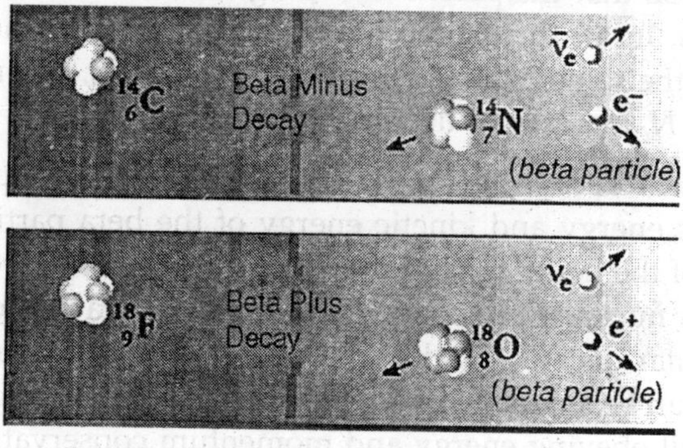

Fig. Beta Decays

These particular reactions take place because conservation laws are obeyed. Electric charge conservation requires that if an electrically neutral neutron becomes a positively charged proton, an electrically negative particle (in this case, an electron) must also be produced. Similarly, conservation of lepton number requires that if a neutron (lepton number = 0) decays into a proton (lepton number = 0) and an electron (lepton number = 1), a particle with a lepton number of -1 (in this case an antineutrino) must also be produced. The leptons emitted in beta decay did not exist in the nucleus before the decay–they are created at the instant of the decay.

To the best of our knowledge, an isolated proton, a hydrogen nucleus with or without an electron, does not decay. However within a nucleus, the beta decay process can change a proton to a neutron. An isolated neutron is unstable and will decay with a half-life of 10.5 minutes. A neutron in a nucleus will decay if a more stable nucleus results; the half-life of the decay depends on the isotope. If it leads to a more stable nucleus, a proton in a nucleus may capture an electron from the atom (electron capture), and change into a neutron and a neutrino.

Proton decay, neutron decay, and electron capture are three ways in which protons can be changed into neutrons or vice-versa; in each decay there is a change in the atomic number, so that the parent and daughter atoms are different elements. In all three processes, the number A of nucleons remains the same, while both proton number, Z, and neutron number, N, increase or decrease by 1.

In beta decay the change in binding energy appears as the mass energy and kinetic energy of the beta particle, the energy of the neutrino, and the kinetic energy of the recoiling daughter nucleus. The energy of an emitted beta particle from a particular decay can take on a range of values because the energy can be shared in many ways among the three particles while still obeying energy and momentum conservation.

Gamma Decay

In gamma decay, a nucleus changes from a higher energy state to a lower energy state through the emission of electromagnetic radiation (photons). The number of protons (and neutrons) in the nucleus does not change in this process, so the parent and daughter atoms are the same chemical element.

Fig. Gamma Decay

In the gamma decay of a nucleus, the emitted photon and recoiling nucleus each have a well-defined energy after the decay. The characteristic energy is divided between only two particles.

The Discovery of Radioactivity

In 1896 Henri Becquerel was using naturally fluorescent minerals to study the properties of x-rays, which had been discovered in 1895 by Wilhelm Roentgen. He exposed potassium uranyl sulfate to sunlight and then placed it on photographic plates wrapped in black paper, believing that the uranium absorbed the sun's energy and then emitted it as x-rays. This hypothesis was disproved on the 26th–27th of February, when his experiment "failed" because it was overcast in Paris. For some reason, Becquerel decided to develop his photographic plates anyway. To his surprise, the images were strong and clear, proving that the uranium emitted radiation without an external source of energy such as the sun. Becquerel had discovered radioactivity.

Becquerel used an apparatus similar to that displayed below to show that the radiation he discovered could not be x-rays. X-rays are neutral and cannot be bent in a magnetic field. The new radiation was bent by the magnetic field so that the radiation must be charged and different than x-rays.

When different radioactive substances were put in the magnetic field, they deflected in different directions or not at all, showing that there were three classes of radioactivity: negative, positive, and electrically neutral.

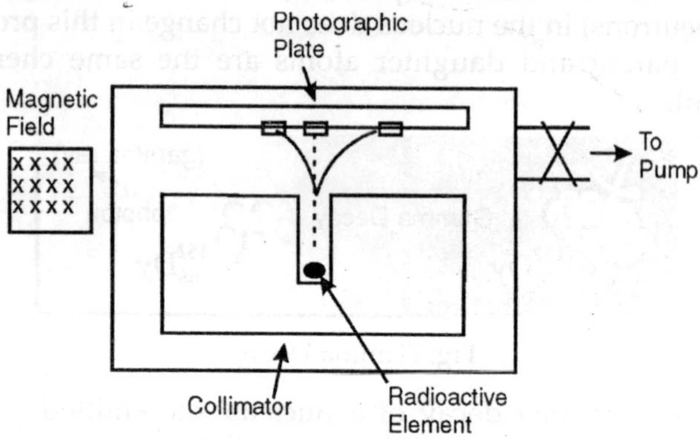

Fig. Radio Activity

The term radioactivity was actually coined by Marie Curie, who together with her husband Pierre, began investigating the phenomenon recently discovered by Becquerel. The Curies extracted uranium from ore and to their surprise, found that the leftover ore showed more activity than the pure uranium. They concluded that the ore contained other radioactive elements. This led to the discoveries of the elements polonium and radium. It took four more years of processing tons of ore to isolate enough of each element to determine their chemical properties.

Ernest Rutherford, who did many experiments studying the properties of radioactive decay, named these alpha, beta, and gamma particles, and classified them by their ability to penetrate matter. When the air from the chamber was removed, the alpha source made a spot on the photographic plate. When air was added, the spot disappeared. Thus, only a few centimeters of air were enough to stop the alpha radiation.

Because alpha particles carry more electric charge, are more massive, and move slowly compared to beta and gamma particles, they interact much more easily with matter.

Beta particles are much less massive and move faster, but are still electrically charged. A sheet of aluminum one millimeter thick or several meters of air will stop these electrons and positrons. Because gamma rays carry no electric charge, they can penetrate large distances through materials before interacting–several centimeters of lead or a meter of concrete is needed to stop most gamma rays.

Radioactivity in Nature

Radioactivity is a natural part of our environment. Present-day Earth contains all the stable chemical elements from the lowest mass (H) to the highest (Pb and Bi). Every element with higher Z than Bi is radioactive. The earth also contains several primordial long-lived radioisotopes that have survived to the present in significant amounts. $_{40}K$, with its 1.3 billion year half-life, has the lowest mass of these isotopes and beta decays to both $_{40}Ar$ and $_{40}Ca$.

Many isotopes can decay by more than one method. When actinium-226 (Z=89) decays, 83percent of the rate is through β --decay, $_{226}Ac \rightarrow {}_{226}Th$ + e- +, 17percent is through electron capture, 226Ac + e- $\rightarrow {}_{226}Fr$ + *n*, and the remainder, 0.006percent, is through α-decay, $_{226}Ac \rightarrow {}_{222}Fr + {}_{4}He$. Therefore from 100,000 atoms of actinium, one would measure on average 83,000 beta particles and 6 alpha particles (plus 100,000 neutrinos or antineutrinos). These proportions are known as branching ratios. The branching ratios are different for the different radioactive nuclei.

Three very massive elements, $_{232}Th$ (14.1 billion year half-life), $_{235}U$ (700 million year half-life), and $_{238}U$ (4.5 billion year half-life) decay through complex "chains" of alpha and beta decays ending at the stable $_{208}Pb$, $_{207}Pb$, and $_{206}Pb$ respectively. The decay chain for $_{238}U$. The ratio of uranium to lead present on Earth today gives us an estimate of its age (4.5 billion years). Given Earth's age, any much shorter lived radioactive nuclei present at its birth have already decayed into stable elements. One of the intermediate products of the ^{238}U decay chain, ^{222}Rn (radon) with a half-life of 3.8 days, is responsible for higher levels of background radiation in many parts of the world. This is primarily because it is a gas and can easily seep out of the earth into unfinished basements and then into the house.

Radioactive isotopes, $_{14}C$ and $_{7}Be$, are produced continuously through reactions of cosmic rays (high energy charged particles from outside Earth) with molecules in the upper atmosphere. $_{14}C$ is useful for radioactive dating. Also, the study of radioactivity is very important to understand the structure of the earth because radioactive decay heats the earth's interior to very high temperatures.

The uranium decay series. The vertical axis is atomic mass.

Units of Radioactivity

The number of decays per second, or activity, from a sample of radioactive nuclei is measured in becquerel (Bq), after Henri Becquerel. One decay per second equals one becquerel.

An older unit is the curie, named after Pierre and Marie Curie. One curie is approximately the activity of 1 gram of radium and equals (exactly) 3.7×10^{10} becquerel. The activity depends only on the number of decays per second, not on the type of decay, the energy of the decay products, or the biological effects of the radiation.

Four Fundamental Interactions

The forces of gravity and electromagnetism are familiar in everyday life. Two new forces are introduced when discussing nuclear phenomena: the strong and weak interactions. When two protons encounter each other, they experience all four of the fundamental forces of nature simultaneously. The weak force governs beta decay and neutrino interactions with nuclei. The strong force, which we generally call the nuclear force, is actually the force that binds quarks together to form baryons (3 quarks) and mesons (a quark and an anti-quark). The nucleons of everyday matter, neutrons and protons, consist of the quark combinations *uud* and *udd*, respectively. The symbol *u* represents a single up quark, while the symbol *d* represents a single down quark.

The force that holds nucleons together to form an atomic nucleus can be thought to be a residual interaction between quarks inside each individual nucleon. This is analogous to what happens in a molecule. The electrons in an atom are bound to its nucleus by electromagnetism: when two atoms are relatively near, there is a residual interaction between the electron clouds that can form a covalent bond. The nucleus can thus be thought of as a "strong force molecule."

The force between two objects can be described as the exchange of a particle. The exchange particle transfers momentum and energy between the two objects, and is said to mediate the interaction. A simple analogue of this is a ball being thrown back and forth between two people. The momentum imparted to the ball by one person gets transferred to the other person when she catches the ball.

Strength and range of the four fundamental forces between two protons. Note that the strong force acts between quarks by an exchange of gluons. The residual strong force between two protons can be described by the exchange of a neutral pion. Note, the W± is not included as an exchange particle for the weak interaction because it is not exchanged in the simplest proton-proton interaction.

The potential energy associated with each force acting between two protons is characterized by both the strength of the interaction and the range over which the interaction takes place. In each case the strength is determined by a coupling constant, and the range is characterized by the mass of the exchanged particle. The potential energy, *U*, between two protons a distance *r* apart is written as

$$U = \frac{C^2}{r}\exp(-r/R),$$

where *R* is the range of the interaction, and C^2, is the strength of the interaction. In each case the interaction is due to the exchange of some particle whose mass determined the range of the interaction, $R = h/mc$. The exchanged particle is said to *mediate* the interaction.

A comparison between the coupling constants and ranges of the four forces acting between two protons. Although the graviton has not yet been observed, it is thought that there is an exchange particle associated with gravity and that eventually gravity will be described in a unified theory with the other three forces of nature.

The range of the gravitational and electromagnetic forces is infinite, while that of the strong and weak forces is very short. Also, the strength of the interaction depends on the separation between the two protons. Both gravitation and electromagnetism are of infinite range and their strengths decrease as the separation, r, increases–falling off as $1/r^2$. On the other hand, because the exchange particles for the strong and weak forces have a large mass, the force associated with them is zero outside of a short range.

Note that the strong force between two protons is a residual interaction. The quarks inside the nucleons interact through the exchange of gluons that carry a quantum number called "colour" (i.e., a "colour-neutral" object does not feel the strong force, like an electrically neutral object does not feel electromagnetism).

The theory of strong interactions among quarks and gluons is called Quantum Chromodynamics (QCD). A very successful model of describing this residual force is by the exchange of the π (pi) meson, or pion. This description, dating back to the work of Yukawa in the 1930s, works extremely well except at very short distances, and is generally used to describe most of the properties of complex nuclei. Exactly at what distance inside the nucleus, or correspondingly at what values of momentum transferred between the nucleons, the pion exchange model breaks down is the subject of much current research in nuclear physics. It is presently thought that this transition occurs somewhere around momentum transfers of 1 GeV/c.

RADIATION IN THE ENVIRONMENT

Many forms of "radiation" are encountered in the natural environment and are produced by modern technology. Most of them have the potential for both beneficial and harmful effects. Even sunlight, the most essential radiation of all, can be harmful in excessive amounts. Most public attention is given to the category of radiation known as "ionizing radiation." This radiation can disrupt atoms, creating positive ions and negative electrons, and cause biological harm. Ionizing radiation includes x-rays, gamma rays, alpha particles, beta particles, neutrons, and the varieties of cosmic rays.

Radiation Damage and its Study

All ionizing radiations, at sufficiently large exposures, can cause cancer. Many, in carefully controlled exposures, are also used for cancer therapy. Whether harmful or beneficial, exposures to ionizing radiation have been an inevitable part of the environment throughout the Earth's history. The nucleosynthesis processes that produced the elements created both stable and unstable nuclides. The unstable nuclides with very long half-lives, together with their radioactive progeny, constitute the natural radioactivity on Earth today. In addition, violent processes in the sun and elsewhere lead to the bombardment of the Earth by cosmic rays. Thus, radiation is an old and familiar, if unrecognized, pollutant.

However, human awareness of radioactivity and ionizing radiation has only a 100-year history starting with the discovery of x-rays and radioactivity. The first evidence that ionizing radiation could do harm came within months after the discovery of x-rays, when an early x-ray worker developed injuries to his skin. Serious efforts to understand and control radiation exposures started in the 1920s and greatly expanded during and after World War II.

Information on the effects of radiation comes from studies of exposed groups and individuals, from animal experiments, and from studies at the cellular and molecular level. It is now well established that ionizing radiation has both prompt and delayed effects. At very high radiation exposures, death will occur within several months or less. At moderate levels, radiation exposure increases the chance that an individual will develop cancer, with a time delay of ten or more years for most cancers. At low levels, the cancer risk decreases, but the relationship between cancer risk and the magnitude of the exposure is uncertain.

Other effects of radiation, in part inferred from animal experiments, include an increased risk of genetic defects and, for exposures of the fetus before birth, of mental retardation. In terms of frequency of occurrence and severity of effects, cancer is the most serious consequence and receives the greatest attention.

The importance of genetic effects has turned out to be much less than was originally expected. In the words of a 1993 NCRP report: "... the genetic risks have been found to be smaller and the cancer risks larger than were thought (in the 1950s)." Strikingly, no statistically significant genetic effects have been found among the extensively studied children of survivors of the Hiroshima-Nagasaki bombings. On the basis of animal experiments, however, one expects some genetic effects–even if so far not observable above the large background of "natural" defects. In addition, the Hiroshima-Nagasaki studies have shown an increased incidence of mental retardation among children who received large prenatal radiation doses, especially for exposures 8 to 15 weeks after conception.

Units of Radioactivity and Dose

The original unit for measuring the amount of radioactivity was the curie (Ci)–first defined to correspond to one gram of radium-226 and more recently defined as:

1 curie = 3.7×10^{10} radioactive decays per second [exactly].

In the International System of Units (SI) the curie has been replaced by the becquerel (Bq), where

1 becquerel = 1 radioactive decay per second = 2.703×10^{-11} Ci.

The magnitude of radiation exposures is specified in terms of the radiation dose. There are two important categories of dose:

The *absorbed dose,* sometimes also known as the *physical dose,* defined by the amount of energy deposited in a unit mass in human tissue or other media. The original unit is the *rad* [100 erg/g]; it is now being widely replaced by the SI unit, the gray (Gy) [1 J/kg], where 1 gray = 100 rad.

- The *biological* dose, sometimes also known as the *dose equivalent,* expressed in units of rem or, in the SI system, *sievert* (Sv). This dose reflects the fact that the biological damage caused by a particle depends not only on the total energy deposited but also on the rate of energy loss per unit distance traversed by the particle (or "linear energy transfer"). Alpha particles do much more damage per unit energy deposited than do electrons. This effect can be represented, in rough overall terms, by a *quality factor, Q*. Over a wide range of incident energies, *Q* is taken to be 1.0 for electrons (and for x-rays and gamma rays, both of which produce electrons) and 20 for alpha particles. For neutrons, the adopted quality factor varies from 5 to 20, depending on neutron energy.

The biological impact is specified by the *dose equivalent* H, which is the product of the absorbed dose D and the quality factor Q: $H = Q\ D$.

The unit for the dose equivalent is the rem if the absorbed dose is in rads and the sievert (Sv) if the absorbed dose is in grays. Thus, 1 Sv = 100 rem. 1 rem is roughly the average dose received in 3 years of exposure to natural radiation. 1 Sv is at the bottom of the range of doses that, if received over a short period of time, are likely to cause noticeable symptoms of radiation sickness.

The dose equivalent is still not the whole story. If only part of the body is irradiated, the dose must be discounted with an appropriate weighting factor if it is to reflect overall risk. The discounted dose is termed the *effective dose equivalent* or just the *effective dose,* expressed in rems or sieverts.

Radioactivity in the Natural Environment

The radioactive nuclei, or radionuclides, found naturally on Earth can be grouped into three series–headed by uranium-238, uranium-235, and thorium-232–plus several isolated beta-particle emitting nuclei, most prominently potassium-40 and rubidium-87.

In the human body the activity of ^{210}Pb and ^{210}Po, both progeny of ^{238}U, is much greater than that of ^{238}U itself.

The most interesting of the series is the uranium-238 series which decays via a chain containing 8 alpha decays and 6 beta decays to lead-206. This chain includes the longest-lived isotopes of radium and radon: radium-226 and radon-222, respectively. In each of the three chains the parent nucleus has a much greater lifetime than does any of the progeny. Therefore, a steady-state is established in which, for a given sample of material, each member of the series has the same activity–aside from deviations due to differences in chemical properties, which cause different elements to be transferred at different rates into or out of a given sample of material.

Including all the succeeding decays, the total activity in the thorium-232 and uranium-238 series is, very roughly, ten times the activity indicated for thorium-232 and uranium-238 alone. Thus, for each of the series, the total activity in the Earth's crust averages roughly 30,000 Ci/km^3. For both series together and including the contributions of potassium-40 and rubidium-87, the total activity in the crust averages about 100,000 Ci/km^3. There is also a considerable amount of radioactivity in the oceans, with potassium-40 dominant in the ocean itself and thorium-232 relatively more important in the ocean sediments. For the oceans as a whole (1.4×10^{21} liters), the total activity is about 4×10^{11} Ci for potassium–40 and 1 ×

10^9 Ci for uranium-238. Potassium–40 is also present in significant amounts in the human body, especially in muscle tissue.

In addition to these ancient radionuclides and their progeny, some radionuclides are being continually produced by cosmic rays. The most prominent of these is carbon-14, produced in the interaction of cosmic ray neutrons with nitrogen in the atmosphere.

Typical Radiation Doses

The chief sources of radiation exposure in the United States, as tabulated by the NCRP. The largest single source of exposure is from radon, which is produced in the decay of radium-226 in the soil and enters a house through openings at the base. The "radon" dose arises mostly from the inhalation of the progeny of radon-222, and varies widely from house-to-house depending upon the radium content of the underlying soil, its porosity, and the house construction. The average effective dose of 2.0 mSv/yr (200 mrem/yr) corresponds to the average radon concentration, but there are more than one million homes with radon levels that are more than five times as great.

Effects of Low Doses

Most of the radiation doses that are received by members of the public and by radiation workers–both routinely and in accidents–are what are commonly referred to as "low doses." There is no precise definition of "low" but it would include doses 10 mSv per year. The average radiation doses received by people in the U.S. are in the "low dose" region. It is obviously important to determine the effects of low radiation doses–or, more precisely, the effects of small additions to the unavoidable natural background dose.

However, despite much study, these effects are not known, being too small to see unambiguously. The most prominent assumption, accepted by most official bodies, is the so-called *linearity hypothesis,* according to which the cancer risk is directly proportional to the magnitude of the dose, down to zero dose. In applying this assumption a consensus estimate is that the risk to a "typical" individual of an eventual fatal cancer is 0.00005 per mSv (or 0.05 per Sv). Thus, if 100,000 people each receive an added dose of 1 mSv, then 5 additional cancer deaths are to be expected. At the same time, while adopting the linearity hypothesis as a prudent working assumption, many of the leading studies have also indicated the possibility that small increases in radiation dose do not create any additional cancer risk. This reflects the considerable disagreement that exists within the scientific community as to the validity of the linearity hypothesis.

Effects of Large Doses

Radiation doses above 3 Gy (300 rad) can be fatal and doses above 6 Gy (600 rad) are almost certain to be fatal, with death occurring within several months (in shorter times at higher doses). Above 1 Gy, radiation causes a complex of symptoms, including nausea and blood changes, known as radiation sickness. For doses below 1 Sv (100 rem), there is little likelihood of radiation sickness, and the main danger is an increased cancer risk. The most important data base and analyses are from the RERF studies of the Hiroshima and Nagasaki survivors. In these studies, the exposure and medical histories are analysed for an exposed group (50,113 people) and an unexposed, or minimally exposed, group (36,459 people). Through 1990, there have been 4,741 cancer fatalities in the exposed group, of which 454 are attributed to radiation exposure. There is a statistically significant excess for both solid cancer tumors and leukemia for doses above 0.2 Sv (20 rem). These data, in a succession of updated versions, have provided much of the information used in comprehensive studies of radiation effects.

Nuclear Reactor Accidents

The accidents at the Three Mile Island (TMI) and Chernobyl nuclear reactors have triggered particularly intense concern about radiation hazards. The TMI accident, in Pennsylvania in 1979, resulted from a combination of deficient equipment and operator errors. Even though there was severe damage to the nuclear fuel within the reactor, very little radioactivity escaped into the outside environment. The effectiveness of the large concrete containment building that surrounded the reactor contributed to this relatively small release. Subsequent studies concluded that the maximum dose received by any member of the public was under 1 mSv (100 mrem). The collective off-site dose is estimated to have been about 20 person-Sv. Under the standard low-dose assumption, this corresponds to one eventual cancer fatality in the neighbouring population of 2 million people. (This population receives an annual collective dose of about 6000 person-Sv from natural sources.)

The 1986 Chernobyl accident was far more serious. It occurred in a reactor with an unsafe reactor design unique to the Soviet Union. The reactor had no effective containment, and there was a very large release of radionuclides to the environment. The accident led to the death within several months of 31 reactor personnel and firefighters–28 from a combination of radiation effects and burns from fire, 2 from other injuries, and one from a heart attack. A total of 237 workers was hospitalized for symptoms of radiation sickness, including the 28 who died. A 1996 summary reported an additional 14 deaths among the more severely exposed workers, but it is not clear that these deaths were all due to the prior exposure.

There is strong evidence of a substantial increase in thyroid cancers among children living in the vicinity. No other health effects from Chernobyl have been convincing established. However, it is too soon for them to have been fully manifested. Standard calculations of radiation effects predict that there will be a large number of excess cancer deaths among the so-called "liquidators," who were engaged in cleanup operations after the accident, as well as in the neighbouring population.

Considering impacts at greater distances, one early study estimated that the collective dose in the Northern Hemisphere over a 50-year period would be about 930,000 person-Sv. While there is substantial uncertainty in the dose estimate, there is even greater uncertainty as to the impact. If one accepts the linearity hypothesis and assumes 0.05 fatalities per Sv, this dose corresponds to 47,000 eventual cancer fatalities. About 29,000 of these fatalities would occur in Europe (outside the former Soviet Union) due to a cumulative collective dose of 580,000 Sv–an average individual *lifetime* dose of 1.2 mSv for 490 million people. Given these low average doses, any estimate of predicted deaths from Chernobyl is highly speculative. The deaths will not be identifiable, being masked by the 88,000,000 "normal" cancer fatalities expected in this region during the 50-year period.

Chapter 13

Nuclear Reactions

Nuclear reactions and nuclear scattering are used to measure the properties of nuclei. Reactions that exchange energy or nucleons can be used to measure the energies of binding and excitation, quantum numbers of energy levels, and transition rates between levels. A particle accelerator produces a beam of high-velocity charged particles (electrons, protons, alphas, or "heavy ions"), which then strikes a target nucleus. Nuclear reactions can also be produced in nature by high-velocity particles from cosmic rays, for instance in the upper atmosphere or in space. Beams of neutrons can be obtained from nuclear reactors or as secondary products when a charged-particle beam knocks out weakly-bound neutrons from a target nucleus. Nuclear reactions can also be produced by beams of photons, mesons, muons, and neutrinos.

In order for a nuclear reaction to occur, the nucleons in the incident particle, or projectile, must interact with the nucleons in the target. Thus the energy must be high enough to overcome the natural electromagnetic repulsion between the protons. This energy "barrier" is called the Coulomb barrier. If the energy is below the barrier, the nuclei will bounce off each other. Early experiments by Rutherford used low-energy alpha particles from naturally radioactive material to bounce off target atoms and measure the size of the target nuclei.

When a collision occurs between the incident particle and a target nucleus, either the beam particle scatters elastically leaving the target nucleus in its ground state or the target nucleus is internally excited and subsequently decays by emitting radiation or nucleons. A nuclear reaction is described by identifying the incident particle, target nucleus, and reaction products. When a neutron strikes a nitrogen nucleus, ^{14}N, to produce a proton, ^{1}H, and an isotope of carbon, ^{14}C, the reaction is written as

$$^{1}_{0}n + ^{14}_{1}N \rightarrow ^{14}_{6}C + ^{1}_{1}H$$

Sometimes the reaction is abbreviated as. A number of conservation conditions apply to any reaction equation:

- The mass number A and the charge Z must balance on each side of the reaction arrow. Thus in the example the sum of superscripts 1 + 14 on the left equals the 14 + 1 on the right to balance the A's. The 0 + 7 subscripts on the left equals the 6 + 1 on the right to balance the Z's.
- The total energy before the reaction must equal the total energy after the reaction. The total energy includes the particle kinetic energies plus the energy equivalent of the particle rest masses, $E = mc^2$.
- Linear momenta before and after the reaction must be equal. For two-particle final states this means that a measurement of one particle's momentum determines the other particle's momentum.
- Quantum rules govern the balancing of the angular momentum, parity, and isospin of the nuclear levels.

A specific reaction is studied by measuring the angles and kinetic energies of the reaction products (the kinematic variables). Particle and radiation detectors designed for the expected charge and energy of each product are arranged around the target.

The most important quantity of interest for a specific set of kinematic variables is the reaction cross section. The cross section is a measure of the probability for a particular reaction to occur. This quantity, σ, which has the dimension of area, is measured by the experimental ratio

$$\sigma = \frac{\text{number of reactions particles emitted}}{\begin{pmatrix}\text{number of beam} \\ \text{particles per unit area}\end{pmatrix}\begin{pmatrix}\text{number of target} \\ \text{nuclei within the beam}\end{pmatrix}}$$

The cross section can also be calculated from a mathematical model of the nucleus by applying the rules of quantum mechanics. Comparing the measured and calculated values of the cross sections for many reactions validates the assumptions of the nuclear model.

Table. Shows some of the many types of nuclear reactions and what they teach us about nuclei and nuclear energy.

The elastic scattering cross sections of protons and neutrons on a proton target give the essential data to reconstruct the nucleon-nucleon. A complete theory of nuclear structure and dynamics must start with this elemental interaction.

The systematics of nuclear sizes, shapes, binding energies, and other nuclear properties are the data that nuclear models are challenged to explain. Most of this data was the result of elastic and inelastic scattering of electrons, protons, and neutrons from nuclei found in the chart of the nuclides.

At high enough excitation energies, a nucleus can undergo a series of normal modes of collective oscillations called giant resonances. The nucleus rings like a bell at distinct frequencies with all the nucleons participating and sharing the excitation energy.

Fusion reactions are the combining of two nuclei to form a more massive nucleus. Many fusion reactions release large amounts of energy. An example is the combining of two isotopes of hydrogen (tritium and deuterium) to form helium and a neutron plus a large amount of kinetic energy in the reaction products:

$$^{3}H + {}^{2}H \rightarrow {}^{4}He + n + 17.6 \text{ MeV}.$$

This reaction as a potential electric power source. Another example of fusion is the reaction set that powers the Sun and other low-mass stars:

$$^{1}H + {}^{1}H \rightarrow {}^{2}H + e+ + n\ e,$$

$$^{2}H + {}^{1}H \rightarrow {}^{3}He + \gamma,$$

$$^{3}He + {}^{3}He \rightarrow {}^{4}He + {}^{1}H + {}^{1}H.$$

The net energy output from this chain is 26.7 MeV for each helium-4 nucleus formed.

Neutron-induced fission of massive nuclei into two lower-mass nuclei plus neutrons is also an energy source for power generation.

Compound nucleus formation is a reaction in which two nuclei combine into a single excited nucleus; the excited nucleus lives for a relatively long time and "forgets" how it was formed. The decay from this state of excitation is by "evapouration" of nucleons from the heated liquid drop of the compound nucleus, by gamma decay, or by fission of the compound nucleus. The statistical nature of this process teaches us about the average properties of excited states of complex nuclei.

Multifragmentation reactions, in which high-energy nuclei collide with other nuclei, are a method of creating nuclear matter in unusual conditions of density and excitation energy. These states may be in a different phase from normal nuclei and be characteristic of the matter in the early universe.

The fundamental force between nucleons in nuclei is dominated by the exchange of π mesons (pions). When these particles are created in high-energy proton reactions, they can be used to bombard nuclear targets. When the pion interacts with a nucleus, it forms a resonance with one of the bound nucleons. The resonance is shifted and broadened compared to the reaction on a free nucleon. These changes reflect the influence of the neighbouring nucleons.

When nucleons are flung at one another, they can mesh briefly. During the time they are one nucleus, the quarks in the nucleons can interact with one another as if they were free particles. As with Rutherford scattering, an investigation of the angle that a particle is scattered gives information about the conditions inside the nucleons.

Nuclear reactions and their interpretation are the main activity of most nuclear scientists. The continuing development of accelerators and detectors permit the refinement of nuclear data and models to benefit basic science and nuclear applications.

ENERGY FROM NUCLEAR SCIENCE

The isotope ^{235}U, with an abundance of only 0.7percent in natural uranium, is commonly used to produce electricity in nuclear fission reactors. This isotope has the distinctive and useful property of undergoing nuclear fission through interaction with thermal-energy neutrons (neutrons with average speeds of only a few km/s). The other main isotope of uranium, ^{238}U, does not undergo nuclear fission with thermal neutrons, but it does capture neutrons to form the isotope ^{239}Np which then decays to ^{239}Pu. This isotope of plutonium undergoes nuclear fission with thermal neutrons with a higher probability than that of ^{235}U. The energy released in the fission of ^{235}U and ^{239}Pu, mainly in the form of kinetic energy of the fission fragments, provides the heat to run the turbines that generate electricity at a nuclear fission power plant.

NUCLEAR FISSION ENERGY

The relevant nuclear reactions can be written as follows:

$^{235}U + {}^{1}n \rightarrow$ fission products + neutrons + energy (~200 MeV) (1)

$^{238}U + {}^{1}n \rightarrow {}^{239}U$ + gamma rays (2)

$^{239}U \rightarrow {}^{239}Np \rightarrow {}^{239}Pu$ (a series of beta decays). (3)

An integer number of neutrons, either two, three or four, are emitted in the reactions leading to different pairs of fission products described in reaction (1). The average number is 2.43. Some of the neutrons in reaction (1) can be used to induce fission in another ^{235}U nucleus, thus continuing a controlled, self-perpetuating nuclear chain reaction. Some fraction of the remaining neutrons from reaction (1) are utilized in reaction (2) to produce ^{239}Pu. The rest are absorbed in other nuclei without further effect.

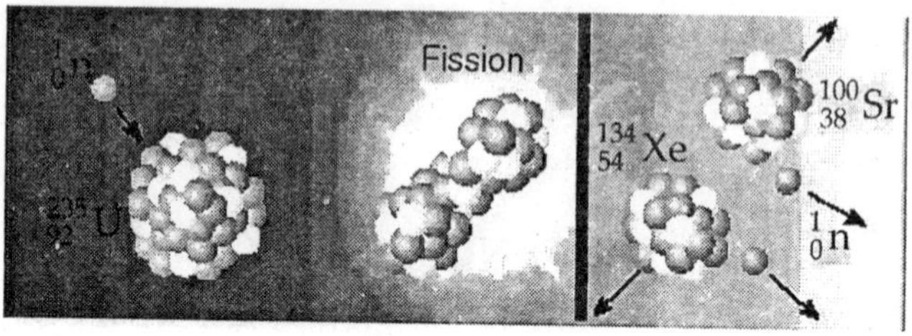

Fig. Fission of ^{235}U after absorption of a thermal neutron.

The isotope ^{232}Th, although not fissionable with thermal neutrons, is a possible energy source because it absorbs thermal neutrons to produce long-lived ^{233}U, which also undergoes thermal neutron fission with a high probability. Thus the "big three" readily fissionable nuclei are: ^{235}U, ^{239}Pu, and ^{233}U.

A typical pressurized (or boiling) water nuclear reactor consists of a core of fissionable material (UO_2, enriched to 3.3percent to 4percent in ^{235}U) in which the chain reaction takes place. The energy released in the fission process, which is primarily in the form of the kinetic energy of the fission fragments, heats the water. The water serves both as a neutron moderator (it slows down the fission neutrons to thermal energies), and as a heat transfer fluid. The chain reaction is controlled by rods of neutron-absorbing material inserted into the core. The thermal energy is removed from the core by the water to an external thermal-energy converter. In the pressurized water reactor (PWR), the thermal energy produces steam for the turbine through the use of a heat exchanger, whereas in a boiling water reactor (BWR), the steam is produced for direct use in the turbine.

Nuclear reactions liberate a large amount of energy compared to chemical reactions. One fission event results in the release of about 200 MeV of energy, or about 3.2×10^{-11}

watt-seconds. Thus, 3.1×10^{10} fissions per second produce 1 W of thermal power. The fission of 1 g of uranium or plutonium per day liberates about 1 MW. This is the energy equivalent of 3 tons of coal or about 600 gallons of fuel oil per day, which when burned produces approximately 1/4 tonne of carbon dioxide. (A tonne, or metric ton, is 1000 kg.)

Nuclear reactors manufacture their own fuel, since they produce ^{239}Pu from ^{238}U. With the total worldwide installed nuclear capacity of 3.4×10^5 MW_e (megawatt electrical), one can estimate that more than 100 tonnes of ^{239}Pu are produced each year in reactors whose primary energy source is the fission of ^{235}U. This ^{239}Pu can be reprocessed from used fuel rods and used to power other reactors.

It is actually possible to generate more ^{239}Pu than is used up in the reactor by surrounding the core with a uranium blanket and generating ^{239}Pu in this blanket. This is called a breeder reactor. A breeder reactor needs to be operated with fast neutrons, a so-called "fast breeder" reactor. In a fast-breeder reactor, water cannot be used as a coolant because it would moderate the neutrons. The smaller fission cross sections associated with the fast neutrons (as compared with thermal neutrons) leads to higher fuel concentrations in the core and higher power densities, which, in turn, create significant heat transfer problems. Liquid sodium metal may be used here as a coolant and heat-transfer fluid. Research on breeder reactors has essentially stopped in the United States because of concerns over nuclear proliferation since the plutonium bred in the reactor might be used for making weapons. Due to such concerns and the complexities of construction and operation, it is unlikely that breeder reactors will ever come into general operation within the next several decades, if ever.

In a yearly operating cycle of a typical (1000 MW_e) pressurized water reactor, the spent fuel contains about 25 tonnes of uranium and about 250 kg of ^{239}Pu. Some 40percent of the energy produced in the course of a nuclear fuel cycle comes from ^{239}Pu. Since about 20percent of the electricity generated in the United States comes from nuclear power plants, about twice as much electricity is generated from ^{239}Pu as is generated from oil-fired electrical generating plants.

Some 435 nuclear power plants operating around the world generate about 345,000 MW_e of electricity in 32 countries, about one-sixth the world's electricity supply. Some countries depend vitally on the electricity generated by nuclear energy. France generates 76percent of its electricity from nuclear power plants; Belgium–56percent, South Korea–36percent, Switzerland–40percent, Sweden–47percent, Finland–30percent, Japan–33percent, and the United Kingdom–25percent. Bulgaria generates 46percent of its electricity from nuclear power, Hungary–42percent, and the Czech Republic and Slovakia combined–20percent. Although the United States is not a leader in percentage, it has the largest total electric output from nuclear power: 98,000 MW_e from 105 plants, generating around 20percent of US electric power.

However, the United States, which led the world in early nuclear electric power development, was also first to be affected by its decline. Over the past two decades the US nuclear electric power industry has received no new domestic orders due to concerns over reactor safety, waste disposal, regulatory uncertainty, increased costs, and decreased electrical demand growth. These same pressures are now affecting nuclear power worldwide, although countries such as France and South Korea still have vigorous programmes. A rebirth may be imminent because of impending solutions to some of these problems and because of the problem of global warming due to the release of greenhouse gases in the combustion of fossil fuels.

The public has become suspicious of nuclear energy partly because of the elaborate methods used to address safety concerns. Most worrisome is the need for external electrical power to supply the pumps used in emergency core cooling systems in the types of reactors now used in the United States. In an accident, this external supply could possibly be disrupted leading to a release of radioactivity. New reactor designs use passive safety to address this problem. Passive safety features can be thought of as characteristics of a reactor that, without operator intervention, will tend to shut or cool a reactor down, keep it in a safe configuration, and prevent release of radioactivity.

These features fall into two broad categories–features that are designed to prevent accidents, and those that mitigate the effects of accidents. Many current problems arise from the huge scale of reactor construction projects. Typically, 1000 MW reactors have been constructed on site. The new types of reactors are smaller (e.g. 600 MW) and may be constructed in factories where uniformity and quality control can produce reactors and operating procedures less prone to failure. Examples of these new designs are two types of Advanced Passive Light Water Reactors (APLWRs), the Liquid Metal Reactor (LMR), and the Modular High Temperature Gas-cooled Reactor (MHTGR). The APLWR is the most developed of these new reactor types. The AP-600 (Advanced Passive 600 MW_e), a pressurized water reactor. It runs at lower temperatures and with larger water inventories than current light water reactors, and has a passive emergency core-cooling system that utilizes gravity. It uses passive natural circulation for the removal of heat from the core.

Nuclear fission reactors, usually pressurized water reactors with energy conversion based on a steam-turbine cycle, have been used extensively to power ships. The United States has built and launched an impressive number of nuclear powered ships–a total of 155 attack and missile submarines, 9 guided missile cruisers (18 reactors), and 5 aircraft carriers (with a total of 24 reactors) and has operated 9 prototype reactors on land. Of this total, 22 submarines have been decommissioned or are non-operational. Shipboard reactors are constructed smaller than similar installations ashore, with special attention given to maintenance, protection from collision, and leakage. Although the primary use of nuclear power has been in submarines and other military vessels, a few non-military marine installations have been demonstrated.

Examples include the Russian icebreaker Lenin and the United States demonstration merchant ship NS Savannah.

Nuclear Fusion Power

Nuclear fusion reactors, if they can be made to work, promise virtually unlimited power for the indefinite future. This is because the fuel, isotopes of hydrogen, are essentially unlimited on Earth. Efforts to control the fusion process and harness it to produce power have been underway in the United States and abroad for more than forty years.

Nuclear fusion is the source of energy in the sun and stars where high temperatures and densities allow the positively-charged nuclei to get close enough to each other for the (attractive) nuclear force to overcome the (repulsive) electrical force and allow fusion to occur. The most promising fusion reaction,

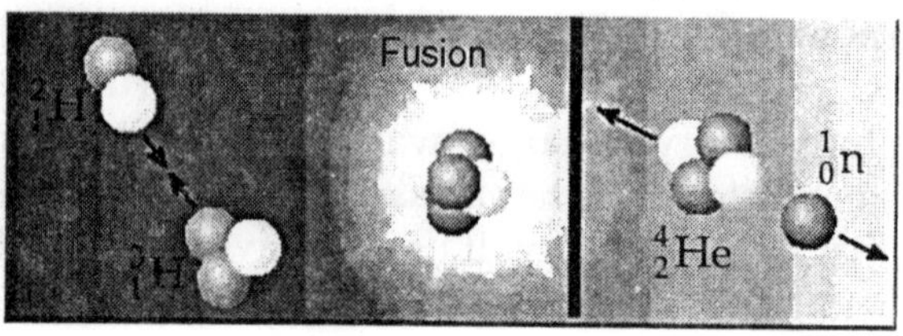

Fig. Nuclear Fusion

^{3}H + ^{2}H Æ ^{4}He + n + 17.6 MeV

involves the radioactive nuclide tritium (^{3}H), available from the nuclear production reaction

^{6}Li + n Æ ^{3}H + ^{4}He.

To produce energy using this reaction, both the magnetic confinement reactor with a high temperature plasma (a gas that has been completely ionized) and the inertial confinement reactor (which utilizes laser implosion technologies) have been investigated. Extremely high plasma temperatures are required in the magnetic confinement reactor and difficult laser implosion techniques are required for the inertial confinement reactor. Although significant progress has been made in these investigations, no working reactor that produces more energy than it consumes has been built. Unfortunately, the funding for continuing this work has declined, and the work is proceeding at a slower pace.

Although these types of reactors would not have the fission product waste disposal problem of fission reactors, fusion reactors generate large number of fast neutrons, leading to large quantities of radioactive byproducts.

Another approach to nuclear fusion–an approach that could lead to aneutronic power (power without neutrons) and non-radioactive nuclear energy–uses the concept of colliding-beam fusion (CBF). One aneutronic method features the ^{2}H + ^{3}He reaction leading to the products ^{1}H + ^{4}He. However, this requires ^{3}He as fuel and terrestrial sources of this are limited. The Moon is a potential source of ^{3}He produced by cosmic-ray protons hitting the Moon directly and not being absorbed by an atmosphere as on Earth. Another potential approach for colliding beam fusion is the ^{11}B + ^{1}H reaction leading to the three ^{4}He nuclei. The energy release is in the form of charged particles whose kinetic energy can be converted to electricity with a very high efficiency. Current research predicts that this energy source has an extremely high degree of cleanness and efficiency. In all current energy sources, approximately two-thirds of the energy is lost in the form of waste heat or thermal pollution. In the CBF approach, there is virtually no waste. This design favours small size for the greatest efficiency (100 MW_e or less), and would lead to either power plants with several reactors or decentralization of energy production.

NUCLEAR FUSION

The deuterium-tritium *(D-T) fusion reaction is considered the m*ost promising for producing sustainable fusion power. From the top: 1. the Deuterium and Tritium nuclei are accelerated towards each other at thermonuclear speeds/ temperatures; 2. they combine to create an unstable helium-5 nucleus; 3. the He-5 nucleus decays, resulting in the ejection of a neutron and repulsion of the He-4 nucleus, both with high energies. For the generation of electrical power by fusion.

In physics and nuclear chemistry, nuclear fusion is the process by which multiple atomic particles join together to form a heavier nucleus. It is accompanied by the release or absorption of energy. Iron and nickel nuclei have the largest binding energies per nucleon of all nuclei and therefore are the most stable. The fusion of two nuclei lighter than iron or nickel generally releases energy while the fusion of nuclei heavier than iron or nickel absorbs energy; vice-versa for the reverse process, nuclear fission.

Nuclear fusion occurs naturally in stars. Artificial fusion in human enterprises has also been achieved, although not yet completely controlled. Building upon the nuclear transmutation experiments of Ernest Rutherford done a few years earlier, fusion of light nuclei (hydrogen isotopes) was first observed by Mark Oliphant in 1932, and the steps of the main cycle of nuclear fusion in stars were subsequently worked out by Hans Bethe throughout the remainder of that decade. Research into fusion for military purposes began in the early 1940s, as part of the Manhattan Project, but was not successful until 1952. Research into controlled fusion for civilian purposes began in the 1950s, and continues to this day.

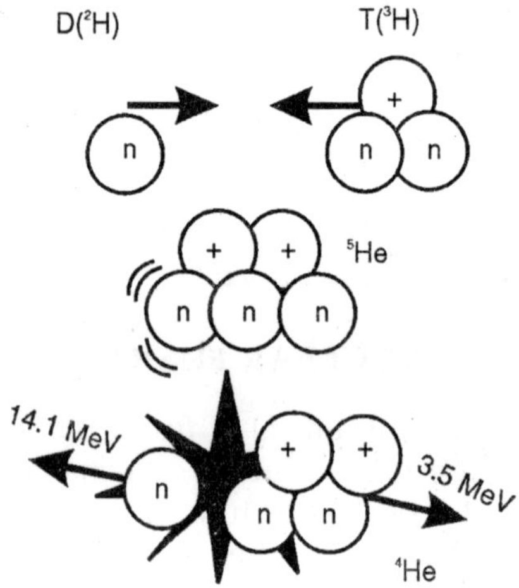

Fig. Fusion reaction

Fusion reactions power the stars and produce all but the lightest elements in a process called nucleosynthesis. Whereas the fusion of light elements in the stars releases energy, production of the heaviest elements absorbs energy, so it can only take place in the extremely high-energy conditions of supernova explosions.

When the fusion reaction is a sustained uncontrolled chain, it can result in a thermonuclear explosion, such as that generated by a hydrogen bomb. Reactions which are not self-sustaining can still release considerable energy, as well as large numbers of neutrons.

Research into controlled fusion, with the aim of producing fusion power for the production of electricity, has been conducted for over 50 years. It has been accompanied by extreme scientific and technological difficulties, and as of yet has not been successful in producing workable designs. As of the present, the only self-sustaining fusion reactions produced by humans have been produced in hydrogen bombs, where the extreme power of a fission bomb is necessary to begin the process. While some plans have been put forth to attempt to use the explosions of hydrogen bombs to generate electricity, none of these have ever moved far past the design stage.

It takes considerable energy to force nuclei to fuse, even those of the lightest element, hydrogen. This is because all nuclei have a positive charge (due to their protons), and as like charges repel, nuclei strongly resist being put too close together. Accelerated to high speeds (that is, heated to thermonuclear temperatures), however, they can overcome this electromagnetic repulsion and get close enough for the strong nuclear force to be active, achieving fusion. The fusion of lighter nuclei, creating a heavier nucleus and a free neutron, will generally release more energy than it took to force them together-an exothermic process that can produce self-sustaining reactions.

The energy released in most nuclear reactions is much larger than that in chemical reactions, because the binding energy that holds a nucleus together is far greater than the energy that holds electrons to a nucleus. The ionization energy gained by adding an electron to a hydrogen nucleus is 13.6

electron volts-less than one-millionth of the 17 MeV released in the D-T (deuterium-tritium) reaction shown to the top right. Fusion reactions have an energy density many times greater than nuclear fission-that is, per unit of mass the reactions produce far greater energies, even though *individual* fission reactions are generally much more energetic than *individual* fusion reactions-which are themselves millions of times more energetic than chemical reactions. Only the direct conversion of mass into energy, such as with collision of matter and antimatter, is more energetic per unit of mass than nuclear fusion.

Requirements

A substantial energy barrier must be overcome before fusion can occur. At large distances two naked nuclei repel one another because of the repulsive electrostatic force between their positively charged protons. If two nuclei can be brought close enough together, however, the electrostatic repulsion can be overcome by the nuclear force which is stronger at close distances.

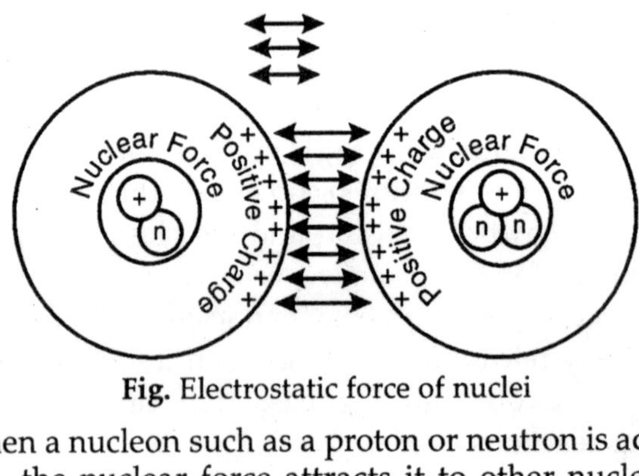

Fig. Electrostatic force of nuclei

When a nucleon such as a proton or neutron is added to a nucleus, the nuclear force attracts it to other nucleons, but primarily to its immediate neighbours due to the short range of the force. The nucleons in the interior of a nucleus have more neighbouring nucleons than those on the surface. Since smaller nuclei have a larger surface area-to-volume ratio, the binding energy per nucleon due to the strong force generally increases with the size of the nucleus but approaches a limiting value corresponding to that of a fully surrounded nucleon.

The electrostatic force, on the other hand, is an inverse-square force, so a proton added to a nucleus will feel an electrostatic repulsion from *all* the other protons in the nucleus. The electrostatic energy per nucleon due to the electrostatic force thus increases without limit as nuclei get larger.

The electrostatic force caused by positively charged nuclei is very strong over long distances, but at short distances the nuclear force is stronger. As such, the main technical difficulty for fusion is getting the nuclei close enough to fuse. Distances not to scale.

The net result of these opposing forces is that the binding energy per nucleon generally increases with increasing size, up to the elements iron and nickel, and then decreases for heavier nuclei. Eventually, the binding energy becomes negative and very heavy nuclei are not stable. The four most tightly bound nuclei, in decreasing order of binding energy, are ^{62}Ni, ^{58}Fe, ^{56}Fe, and ^{60}Ni. Even though the nickel isotope, ^{62}Ni, is more stable, the iron isotope ^{56}Fe is an order of magnitude more common. This is due to a greater disintegration rate for ^{62}Ni in the interior of stars driven by photon absorption.

A notable exception to this general trend is the helium-4 nucleus, whose binding energy is higher than that of lithium, the next heavier element. The Pauli exclusion principle provides an explanation for this exceptional behaviour — it says that because protons and neutrons are fermions, they cannot exist in exactly the same state. Each proton or neutron energy state in a nucleus can accommodate both a spin up particle and a spin down particle. Helium-4 has an anomalously large binding energy because its nucleus consists of two protons and two neutrons; so all four of its nucleons can be in the ground state. Any additional nucleons would have to go into higher energy states.

The situation is similar if two nuclei are brought together. As they approach each other, all the protons in one nucleus repel all the protons in the other. Not until the two nuclei actually come in contact can the strong nuclear force take over. Consequently, even when the final energy state is lower, there is a large energy barrier that must first be overcome. It is called the Coulomb barrier.

The Coulomb barrier is smallest for isotopes of hydrogen—they contain only a single positive charge in the nucleus. A bi-proton is not stable, so neutrons must also be involved, ideally in such a way that a helium nucleus, with its extremely tight binding, is one of the products.

Using deuterium-tritium fuel, the resulting energy barrier is about 0.01 MeV. In comparison, the energy needed to remove an electron from hydrogen is 13.6 eV, about 750 times less energy. The (intermediate) result of the fusion is an unstable ^{5}He nucleus, which immediately ejects a neutron with 14.1 MeV. The recoil energy of the remaining ^{4}He nucleus is 3.5 MeV, so the total energy liberated is 17.6 MeV. This is many times more than what was needed to overcome the energy barrier.

If the energy to initiate the reaction comes from accelerating one of the nuclei, the process is called *beam-target* fusion; if both nuclei are accelerated, it is *beam-beam* fusion. If the nuclei are part of a plasma near thermal equilibrium, one speaks of *thermonuclear* fusion. Temperature is a measure of the average kinetic energy of particles, so by heating the nuclei they will gain energy and eventually have enough to overcome this 0.01 MeV. Converting the units between electronvolts and kelvins shows that the barrier would be overcome at a temperature in excess of 120 million kelvins, obviously a very high temperature.

There are two effects that lower the actual temperature needed. One is the fact that temperature is the *average* kinetic energy, implying that some nuclei at this temperature would actually have much higher energy than 0.01 MeV, while others would be much lower. It is the nuclei in the high-energy tail of the velocity distribution that account for most of the fusion reactions. The other effect is quantum tunneling. The nuclei do not actually have to have enough energy to overcome the Coulomb barrier completely. If they have nearly enough energy, they can tunnel through the remaining barrier. For this reason fuel at lower temperatures will still undergo fusion events, at a lower rate.

The fusion reaction rate increases rapidly with temperature until it maximizes and then gradually drops off. The DT rate peaks at a lower temperature (about 70 keV, or 800 million kelvins) and at a higher value than other reactions commonly considered for fusion energy.

The reaction cross section σ is a measure of the probability of a fusion reaction as a function of the relative velocity of the two reactant nuclei. If the reactants have a distribution of velocities, e.g. a thermal distribution with thermonuclear fusion, then it is useful to perform an average over the distributions of the product of cross section and velocity. The reaction rate (fusions per volume per time) is $<\sigma v>$ times the product of the reactant number densities:

$$f = n_1 n_2 \langle \sigma v \rangle$$

If a species of nuclei is reacting with itself, such as the DD reaction, then the product $n_1 n_2$ must be replaced by $(1 / 2)n^2$.

$\langle \sigma v \rangle$ increases from virtually zero at room temperatures up to meaningful magnitudes at temperatures of 10 – 100 keV. At these temperatures, well above typical ionization energies (13.6 eV in the hydrogen case), the fusion reactants exist in a plasma state.

The significance of $\langle \sigma v \rangle$ as a function of temperature in a device with a particular energy confinement time is found by considering the Lawson criterion.

FUEL CONFINEMENT METHODS

Gravitational

One force capable of confining the fuel well enough to satisfy the Lawson criterion is gravity. The mass needed, however, is so great that gravitational confinement is only found in stars (the smallest of which are brown dwarfs). Even if the more reactive fuel deuterium were used, a mass greater than that of the planet Jupiter would be needed.

Magnetic

The charged ions of fusion fuel follow spiral orbits around magnetic field lines, and the fuel is therefore trapped along the field lines. A variety of magnetic configurations exist, including the toroidal geometries of tokamaks and stellarators and open ended mirror confinement systems.

Inertial

A third confinement principle is to apply a rapid pulse of energy to a large part of the surface of a pellet of fusion fuel, causing it to simultaneously "implode" and heat to very high pressure and temperature. If the fuel is dense enough and hot enough, the fusion reaction rate will be high enough to burn a significant fraction of the fuel before it has dissipated. To achieve these extreme conditions, the initially cold fuel must be explosively compressed. Inertial confinement is used in the hydrogen bomb, where the driver is x-rays created by a fission bomb. Inertial confinement is also attempted in "controlled" nuclear fusion, where the driver is a laser, ion, or electron beam, or a Z-pinch.

Some other confinement principles have been investigated, such as muon-catalyzed fusion, the Farnsworth-Hirsch fusor and Polywell (inertial electrostatic confinement), and bubble fusion.

PRODUCTION METHODS

A variety of methods are known to effect nuclear fusion. Some are "cold" in the strict sense that no part of the material is hot (except for the reaction products), some are "cold" in the limited sense that the bulk of the material is at a relatively low temperature and pressure but the reactants are not, and some are "hot" fusion methods that create macroscopic regions of very high temperature and pressure.

Locally Cold Fusion

- Muon-catalyzed fusion is a well-established and reproducible fusion process that occurs at ordinary temperatures. It was studied in detail by Steven Jones in the early 1980s. It has not been reported to produce net energy. Net energy production from this reaction is not believed to be possible because of the energy required to create muons, their 2.2 μs half-life, and the chance that a muon will bind to the new alpha particle and thus stop catalyzing fusion.
- "Cold fusion" also refers to a simple method using electrodes in heavy water, which has not been verified to be possible.

Generally Cold, locally Hot Fusion

- Accelerator based light-ion fusion. Using particle accelerators it is possible to achieve particle kinetic energies sufficient to induce many light ion fusion reactions. Of particular relevance into this discussion are devices referred to as sealed-tube neutron generators. These small devices are miniature particle accelerators filled with deuterium and tritium gas in an arrangement which allows ions of these nuclei to be accelerated against hydride targets, also containing deuterium and tritium, where fusion takes place. Hundreds of neutron generators are produced annually for use in the petroleum industry where they are used in measurement equipment for locating and mapping oil reserves. Despite periodic reports in the popular press by scientists claiming to have invented "table-top" fusion machines, neutron generators have been around for half a century. The sizes of these devices vary but the smallest instruments are often packaged in sizes smaller than a loaf of bread. These devices do not produce a net power output.
- In sonoluminescence, acoustic shock waves create temporary bubbles that collapse shortly after

creation, producing very high temperatures and pressures. In 2002, Rusi P. Taleyarkhan reported the possibility that bubble fusion occurs in those collapsing bubbles (aka sonofusion). As of 2005, experiments to determine whether fusion is occurring give conflicting results. If fusion is occurring, it is because the local temperature and pressure are sufficiently high to produce hot fusion. In an episode of Horizon, on BBC television, it was conclusively shown that, although temperatures were reached which could initiate fusion on a large scale, no fusion was occurring, and inaccuracies in the measuring system were the cause of anomalous results.

- The Farnsworth-Hirsch Fusor is a tabletop device in which fusion occurs. This fusion comes from high effective temperatures produced by electrostatic acceleration of ions. The device can be built inexpensively, but it too is unable to produce a net power output.
- Antimatter-initialized fusion uses small amounts of antimatter to trigger a tiny fusion explosion. This has been studied primarily in the context of making nuclear pulse propulsion feasible. This is not near becoming a practical power source, due to the cost of manufacturing antimatter alone.
- Pyroelectric fusion was reported in April 2005 by a team at UCLA. The scientists used a pyroelectric crystal heated from –34 to 7°C (–30 to 45°F), combined with a tungsten needle to produce an electric field of about 25 gigavolts per meter to ionize and accelerate deuterium nuclei into an erbium deuteride target. Though the energy of the deuterium ions generated by the crystal has not been directly measured, the authors used 100 keV (a temperature of about 10^9 K) as an estimate in their modeling..At these energy levels, two deuterium nuclei can fuse together to produce a helium-3 nucleus, a 2.45 MeV

neutron and bremsstrahlung. Although it makes a useful neutron generator, the apparatus is not intended for power generation since it requires far more energy than it produces.

Hot Fusion

In "standard" "hot" fusion, the fuel reaches tremendous temperature and pressure inside a fusion reactor or nuclear weapon.

The methods in the second group are examples of non-equilibrium systems, in which very high temperatures and pressures are produced in a relatively small region adjacent to material of much lower temperature. In his doctoral thesis for MIT, Todd Rider did a theoretical study of all quasineutral, isotropic, non-equilibrium fusion systems. He demonstrated that all such systems will leak energy at a rapid rate due to bremsstrahlung radiation produced when electrons in the plasma hit other electrons or ions at a cooler temperature and suddenly decelerate. The problem is not as pronounced in a hot plasma because the range of temperatures, and thus the magnitude of the deceleration, is much lower. Note that Rider's work does not apply to non-neutral and/or anisotropic non-equilibrium plasmas.

IMPORTANT REACTIONS

Astrophysical Reaction Chains

The proton-proton chain dominates in stars the size of the Sun or smaller.

The most important fusion process in nature is that which powers the stars. The net result is the fusion of four protons into one alpha particle, with the release of two positrons, two neutrinos (which changes two of the protons into neutrons), and energy, but several individual reactions are involved, depending on the mass of the star. For stars the size of the sun or smaller, the proton-proton chain dominates. In heavier stars, the CNO cycle is more important. Both types of processes are responsible for the creation of new elements as part of stellar nucleosynthesis.

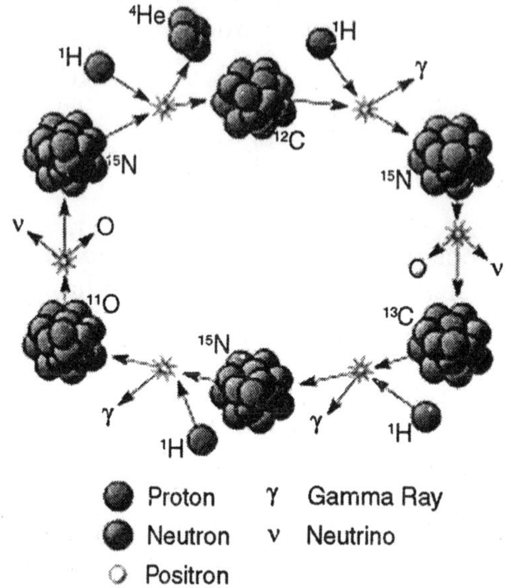

Fig. The CNO cycle dominates in stars heavier than the Sun.

At the temperatures and densities in stellar cores the rates of fusion reactions are notoriously slow. At solar core temperature (T ≈ 15 MK) and density (120 g/cm³), the energy release rate is only 276 mW/cm³—about a quarter of the volumetric rate at which a resting human body generates heat. Thus, reproduction of stellar core conditions in a lab for nuclear fusion power production is completely impractical. Because nuclear reaction rates strongly depend on temperature ($\exp(-E/kT)$), then in order to achieve reasonable rates of energy production in terrestrial fusion reactors 10–100 times higher temperatures (compared to stellar interiors) are required T ≈ 0.1–1.0 GK.

Criteria and Candidates for Terrestrial Reactions

In man-made fusion, the primary fuel is not constrained to be protons and higher temperatures can be used, so reactions with larger cross-sections are chosen. This implies a lower Lawson criterion, and therefore less startup effort. Another concern is the production of neutrons, which activate the reactor structure radiologically, but also have the advantages of allowing volumetric extraction of the fusion energy and tritium breeding. Reactions that release no neutrons are referred to as *aneutronic*.

In order to be useful as a source of energy, a fusion reaction must satisfy several criteria. It must

- Be exothermic: This may be obvious, but it limits the reactants to the low Z (number of protons) side of the curve of binding energy. It also makes helium He-4 the most common product because of its extraordinarily tight binding, although He-3 and H-3 also show up;
- Involve low Z nuclei: This is because the electrostatic repulsion must be overcome before the nuclei are close enough to fuse;
- Have two reactants: At anything less than stellar densities, three body collisions are too improbable. It should be noted that in inertial confinement, both stellar densities and temperatures are exceeded to compensate for the shortcomings of the third parameter of the Lawson criterion, ICF's very short confinement time;
- Have two or more products: This allows simultaneous conservation of energy and momentum without relying on the electromagnetic force;
- Conserve both protons and neutrons: The cross sections for the weak interaction are too small.

For reactions with two products, the energy is divided between them in inverse proportion to their masses, as shown. In most reactions with three products, the distribution of energy varies. For reactions that can result in more than one set of products, the branching ratios are given.

Some reaction candidates can be eliminated at once. The D-^{6}Li reaction has no advantage compared to p-^{11}B because it is roughly as difficult to burn but produces substantially more neutrons through D-D side reactions. There is also a p-^{7}Li reaction, but the cross section is far too low, except possibly when $T_i > 1$ MeV, but at such high temperatures an endothermic, direct neutron-producing reaction also becomes very significant. Finally there is also a p-^{9}Be reaction, which is not only difficult to burn, but ^{9}Be can be easily induced to split into two alphas and a neutron.

In addition to the fusion reactions, the following reactions with neutrons are important in order to "breed" tritium in "dry" fusion bombs and some proposed fusion reactors:

$$n + {}^6Li \rightarrow T + {}^4He$$

$$n + {}^7Li \rightarrow T + {}^4He + n$$

To evaluate the usefulness of these reactions, in addition to the reactants, the products, and the energy released, one needs to know something about the cross section. Any given fusion device will have a maximum plasma pressure that it can sustain, and an economical device will always operate near this maximum. Given this pressure, the largest fusion output is obtained when the temperature is chosen so that $\langle \sigma v \rangle / T^2$ is a maximum.

This is also the temperature at which the value of the triple product $nT\hat{o}$ required for ignition is a minimum, since that required value is inversely proportional to $\langle \sigma v \rangle / T^2$. (A plasma is "ignited" if the fusion reactions produce enough power to maintain the temperature without external heating.) This optimum temperature and the value of $\langle \sigma v \rangle / T^2$ at that temperature is given for a few of these reactions in the following table.

fuel	*T* [keV]	[m^3/s/keV^2]
D-T	13.6	1.24×10^{-24}
D-D	15	1.28×10^{-26}
D-^{3}He	58	2.24×10^{-26}
p-^{6}Li	66	1.46×10^{-27}
p-^{11}B	123	3.01×10^{-27}

Note that many of the reactions form chains. For instance, a reactor fueled with T and ^3He will create some D, which is then possible to use in the D + ^3He reaction if the energies are "right". The ^3He from reaction can react with ^6Li in reaction before completely thermalizing. This produces an energetic proton which in turn undergoes reaction before thermalizing. A detailed analysis shows that this idea will not really work well, but it is a good example of a case where the usual assumption of a Maxwellian plasma is not appropriate.

Neutronicity, Confinement Requirement, and Power Density

The only fusion reactions thus far produced by humans to achieve ignition are those which have been created in hydrogen bombs; the first of which, Ivy Mike, is shown here.

Any of the reactions above can in principle be the basis of fusion power production. In addition to the temperature and cross section discussed above, we must consider the total energy of the fusion products E_{fus}, the energy of the charged fusion products E_{ch}, and the atomic number Z of the non-hydrogenic reactant.

Specification of the D-D reaction entails some difficulties, though. To begin with, one must average over the two branches. More difficult is to decide how to treat the T and ^3He products. T burns so well in a deuterium plasma that it is almost impossible to extract from the plasma. The D-^3He reaction is optimized at a much higher temperature, so the burnup at the optimum D-D temperature may be low, so it seems reasonable to assume the T but not the ^3He gets burned up and adds its energy to the net reaction. Thus we will count the DD fusion energy as E_{fus} = (4.03+17.6+3.27)/2 = 12.5 MeV and the energy in charged particles as E_{ch} = (4.03+3.5+0.82)/2 = 4.2 MeV.

Another unique aspect of the D-D reaction is that there is only one reactant, which must be taken into account when calculating the reaction rate.

With this choice, we tabulate parameters for four of the most important reactions.

fuel	Z	E_{fus} [MeV]	E_{ch} [MeV]	neutronicity
D-T	1	17.6	3.5	0.80
D-D	1	12.5	4.2	0.66
D-^{3}He	2	18.3	18.3	~0.05
p-^{11}B	5	8.7	8.7	~0.001

The last column is the neutronicity of the reaction, the fraction of the fusion energy released as neutrons. This is an important indicator of the magnitude of the problems associated with neutrons like radiation damage, biological shielding, remote handling, and safety. For the first two reactions it is calculated as $(E_{fus}-E_{ch})/E_{fus}$. For the last two reactions, where this calculation would give zero, the values quoted are rough estimates based on side reactions that produce neutrons in a plasma in thermal equilibrium.

Of course, the reactants should also be mixed in the optimal proportions. This is the case when each reactant ion plus its associated electrons accounts for half the pressure. Assuming that the total pressure is fixed, this means that density of the non-hydrogenic ion is smaller than that of the hydrogenic ion by a factor 2/(Z+1). Therefore the rate for these reactions is reduced by the same factor, on top of any differences in the values of $\langle \sigma v \rangle > T^2$. On the other hand, because the D-D reaction has only one reactant, the rate is twice as high as if the fuel were divided between two hydrogenic species.

Thus there is a "penalty" of (2/(Z+1)) for non-hydrogenic fuels arising from the fact that they require more electrons, which take up pressure without participating in the fusion reaction. (It is usually a good assumption that the electron temperature will be nearly equal to the ion temperature. Some authors, however discuss the possibility that the electrons could be maintained substantially colder than the ions. In such a case, known as a "hot ion mode", the "penalty" would not apply.) There is at the same time a "bonus" of a factor 2 for D-D due to the fact that each ion can react with any of the other ions, not just a fraction of them.

We can now compare these reactions in the following table.

fuel	$\langle\sigma v\rangle/T^2$	penalty/ bonus	reactivity	Lawson criterion	power density
D-T	1.24×10^{-24}	1	1	1	1
D-D	1.28×10^{-26}	2	48	30	68
D-^{3}He	2.24×10^{-26}	2/3	83	16	80
p-^{11}B	3.01×10^{-27}	1/3	1240	500	2500

The maximum value of $\langle\sigma v\rangle > T^2$ is taken from a previous table. The "penalty/bonus" factor is that related to a non-hydrogenic reactant or a single-species reaction. The values in the column "reactivity" are found by dividing 1.24 $\times$ 10^{-24} by the product of the second and third columns. It indicates the factor by which the other reactions occur more slowly than the D-T reaction under comparable conditions. The column "Lawson criterion" weights these results with E_{ch} and gives an indication of how much more difficult it is to achieve ignition with these reactions, relative to the difficulty for the D-T reaction. The last column is labeled "power density" and weights the practical reactivity with E_{fus}. It indicates how much lower the fusion power density of the other reactions is compared to the D-T reaction and can be considered a measure of the economic potential.

Bremsstrahlung Losses in Quasineutral, Isotropic Plasmas

The ions undergoing fusion in many systems will essentially never occur alone but will be mixed with electrons that in aggregate neutralize the ions' bulk electrical charge and form a plasma. The electrons will generally have a temperature comparable to or greater than that of the ions, so they will collide with the ions and emit x-ray radiation of 10-30 keV energy (Bremsstrahlung). The Sun and stars are opaque to x-rays, but essentially any terrestrial fusion reactor will be optically thin for x-rays of this energy range. x-rays are difficult to reflect but they are effectively absorbed (and converted into heat) in less than mm thickness of stainless steel (which is part of a reactor's shield). The ratio of fusion power produced to x-ray radiation lost to walls is an important figure of merit. This ratio is generally maximized at a much higher temperature than that which maximizes the power density. The following table shows the rough optimum temperature and the power ratio at that temperature for several reactions.

The actual ratios of fusion to Bremsstrahlung power will likely be significantly lower for several reasons. For one, the calculation assumes that the energy of the fusion products is transmitted completely to the fuel ions, which then lose energy to the electrons by collisions, which in turn lose energy by Bremsstrahlung. However because the fusion products move much faster than the fuel ions, they will give up a significant fraction of their energy directly to the electrons. Secondly, the plasma is assumed to be composed purely of fuel ions. In practice, there will be a significant proportion of impurity ions, which will lower the ratio. In particular, the fusion products themselves *must* remain in the plasma until they have given up their energy, and *will* remain some time after that in any proposed confinement scheme. Finally, all channels of energy loss other than Bremsstrahlung have been neglected. The last two factors are related. On theoretical and experimental grounds, particle and energy confinement seem to be closely related. In a confinement scheme that does a good job of retaining energy, fusion products will build up. If the fusion products are efficiently ejected, then energy confinement will be poor, too.

The temperatures maximizing the fusion power compared to the Bremsstrahlung are in every case higher than the temperature that maximizes the power density and minimizes the required value of the fusion triple product. This will not change the optimum operating point for D-T very much because the Bremsstrahlung fraction is low, but it will push the other fuels into regimes where the power density relative to D-T is even lower and the required confinement even more difficult to achieve. For D-D and D-^{3}He, Bremsstrahlung losses will be a serious, possibly prohibitive problem. For ^{3}He-^{3}He, p-^{6}Li and p-^{11}B the Bremsstrahlung losses appear to make a fusion reactor using these fuels with a quasineutral, anisotropic plasma impossible. This limitation does not apply to non-neutral and anisotropic plasmas; however, these have their own challenges to contend with.

Advantages and Disadvantages

Advantages

- The fuel for fusion reactions are readily available. Deuterium and Tritium are virtually inexhaustible.
- Unlike the burning of coal or other fossil fuels, fusion does not emit harmful toxins into the atmosphere. The combustion of most fossil fuels involves some form of the reaction

  $$C + O_2(g) \rightarrow CO_2(g) + \text{heat}$$

- The carbon dioxide ($CO_2(g)$) emitted by this reaction contributes to the global warning/so-called "Greenhouse Effect" that we've all heard so much about. Fusion, however, produces only helium, a gas that is already in abundance in the atmosphere and will not contribute to global warming.
- A major concern with the use of fission power is the issue of nuclear waste, a dangerous material that can both directly injure people and be manufactured into weapons. Fusion has no such problems with dangerous by-products.

Disadvantages

- Scientists have not yet been able to contain a fusion reaction long enough for there to be a net energy gain.
- Many countries are phasing out fusion research because of the failure to reach a breakthrough.

THEORY

The theory behind fusion is actually relatively simple. Fusion has occurred naturally for billions of years from the farthest reaches of the universe to our very own Milky Way. Simply put, fusion is the process that powers the stars.

The energy that comes from the process relies on the joining, or "fusing," of two atoms to form a new molecule. When this larger, relatively unstable molecule splits apart, it releases energy. The most common fusion process occurs with tritium and deuterium as its fuel. Tritium and deuterium are just big names for different isotopes of hydrogen.

So we know that tritium and deuterium fuse and then split apart to create energy. But where does the energy come from? First, let's look at the actual process of fusion. As we all now know, tritium atoms have 1 proton and 2 neutrons, and deuterium atoms have 1 proton and 1 neutron. So when a deuterium and a tritium bang together, or "fuse," a molecule with 2 protons and 3 neutrons is formed. This molecule is inherently unstable, so it splits apart. But it doesn't just revert to a tritium and deuterium. Instead, it splits into an extremely stable helium atom (2 neutrons and 2 protons). The extra neutron is simply released altogether.

Fig. Isotopes of Hydrogen

There are two explanations for the production of energy, one general and one mathematical. The general explanation is that because the process results in the formation of helium, and because helium is more stable than deuterium and tritium, the system has gone from a state of high potential energy to low potential energy, and thus kinetic energy is given off.

A more mathematical explanation is derived from Albert Einstein's famous equation, $E=mc^2$. Otherwise, just accept the fact that the total mass of a helium atom and a neutron (the products of the reaction) is less than the total mass of a deuterium atom and a tritium atom (the reactants). The "missing mass" has in fact been converted to energy based on Einstein's equation.

The Availability of Fusion Power

The future of fusion power, tritium and deuterium are virtually inexhaustible resources. There is 1 deuterium atom in every 6500 hydrogen atoms in ordinary seawater. Tritium, the other reactant, can be bred from lithium, an element common in the earth's crust.

Thus, theoretically, humans can continue to utilize fusion power indefinitely. After all, the amount of seawater on earth is practically unlimited.

Although fusion power sounds extremely attractive right now, it still has not been developed into a viable energy source. The reason is because it isn't as simple as it sounds. Tritium and deuterium atoms don't (thank goodness) randomly collide and give off energy. They must be heated up to extremely high temperatures (around 100 million degrees!) in order for fusion to take place.

So why can't we just heat them up? Well, when matter is present at the high temperatures necessary for fusion, it changes into another form. We're all familiar with the 3 main states of matter: solid, liquid, and gas. Each succeeding state is more energetic than the last. The atoms of a substance in the gaseous state move about much more than those of a substance in the solid state. Plasma, what has been called the "fourth state of matter," consists of a cloud of charged particles and is the most energetic of the four states of matter. Plasma is the most common form of matter in the universe. In fact, stars are made mostly of plasma. Examples of this unique state of matter here on earth include lightning, neon signs, and ordinary flames.

Because plasma exists only at such high temperatures, the electrons in the atoms of a substance in the plasmid state have enough potential energy to "break free" of the nucleus. Thus, the plasma is said to be made up of a cloud of positively and negatively charged particles. The positive ions are the atoms minus the electrons (and thus have a net positive charge). The negative particles are of course the electrons themselves that have been stripped from the atoms.

One characteristic of matter in the plasma state is how difficult it is to confine. Naturally, anything present at such high temperatures is very energetic and therefore moves around a great deal, and anything that moves around a lot is hard to confine. The containment problem has become one of the biggest setbacks to the use of fusion power on earth.

Methods of Containing Plasma

There are currently three known methods of confining and heating plasma.

The first method, gravity, is the way stars throughout the galaxy keep the plasma confined long enough to produce significant energy outputs. The compression of the plasma by gravitational forces and the subsequent fusion energy produced keep the system heated and continue fusion reactions. The plasma confined by this system lasts upwards of 30 million years. However, scientists here on earth are not yet able to recreate the amazing confining forces characteristic of gravity-bound fusion systems. the extremely famous photograph of M16, a newly forming star system.

The second method, inertia, utilizes laser beams to compress the plasma. Unfortunately, this method requires extreme precision (the entire system is less than 10 cm), and the plasma produced by these means typically lasts for less than 1 tenth of a nanosecond (10^{-7} seconds), an extremely brief time. A piece of equipment at the forefront of inertia-driven fusion reactions is the Glass Nova Laser, currently housed at The Lawrence Livermore National Laboratory (LLNL).

The third method utilizes magnetic fields to compress the plasma. An electromagnetic field encapsulates the plasma chamber with currents that flow in the shape of a helix. The "tokamak" is a type of magnetic field confinement chamber which today is seen as having one of the most promising prospects for harnessing fusion energy.

NUCLEAR FISSION

Nuclear fission is the splitting of the nucleus of an atom into parts (lighter nuclei) often producing photons (in the form of gamma rays), free neutrons and other subatomic particles as by-products. Fission of heavy elements is an exothermic reaction which can release large amounts of energy both as electromagnetic radiation and as kinetic energy of the fragments (heating the bulk material where fission takes place). Fission is a form of elemental transmutation because the resulting fragments are not the same element as the original atom.

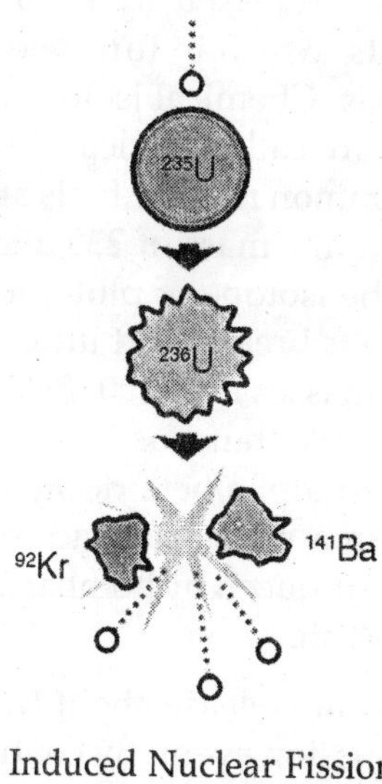

Fig. An Induced Nuclear Fission Event.

Nuclear fission produces energy for nuclear power and to drive the explosion of nuclear weapons. Both uses are made possible because certain substances called nuclear fuels undergo fission when struck by free neutrons and in turn generate neutrons when they break apart. This makes possible a self-sustaining chain reaction that releases energy at a controlled rate in a nuclear reactor or at a very rapid uncontrolled rate in a nuclear weapon. The amount of free energy contained in nuclear fuel is millions of times the amount of free energy contained in a similar mass of chemical fuel such as gasoline, making nuclear fission a very tempting source of energy; however, the products of nuclear fission are radioactive and remain so for significant amounts of time, giving rise to a nuclear waste problem. Concerns over nuclear waste accumulation and over the destructive potential of nuclear weapons may counterbalance the desirable qualities of fission as an energy source, and give rise to ongoing political debate over nuclear power.

PHYSICAL OVERVIEW

Nuclear fission differs from other forms of radioactive decay in that it can be harnessed and controlled via a chain reaction: free neutrons released by each fission event can trigger yet more events, which in turn release more neutrons and cause more fissions. Chemical isotopes that can sustain a fission chain reaction are called nuclear fuels, and are said to be fissile. The most common nuclear fuels are ^{235}U (the isotope of uranium with an atomic mass of 235 and of use in nuclear reactors) and ^{239}Pu (the isotope of plutonium with an atomic mass of 239). These fuels break apart into a range of chemical elements with atomic masses near 100 (fission products). Most nuclear fuels undergo spontaneous fission only very slowly, decaying mainly via an alpha/beta decay chain over periods of millennia to eons. In a nuclear reactor or nuclear weapon, most fission events are induced by bombardment with another particle such as a neutron.

Typical fission events release about two hundred million eV of energy for each fission event. By contrast, most chemical oxidation reactions (such as burning coal or TNT) release at most a few eV per event, so nuclear fuel contains at least ten million times more usable energy than does chemical fuel. The energy of nuclear fission is released as kinetic energy of the fission products and fragments, and as electromagnetic radiation in the form of gamma rays; in a nuclear reactor, the energy is converted to heat as the particles and gamma rays collide with the atoms that make up the reactor and its working fluid, usually water or occasionally heavy water.

Nuclear fission of heavy elements produces energy because the specific binding energy (binding energy per mass) of intermediate-mass nuclei with atomic numbers and atomic masses close to ^{61}Ni and ^{56}Fe is greater than the specific binding energy of very heavy nuclei, so that energy is released when heavy nuclei are broken apart.

The total rest masses of the fission products (Mp) from a single reaction is less than the mass of the original fuel nucleus (M). The excess mass Äm = M - Mp is the invariant mass of the energy that is released as photons (gamma rays) and kinetic energy of the fission fragments, according to the mass-energy equivalence formula $E = mc^2$.

In nuclear fission events the nuclei may break into any combination of lighter nuclei, but the most common event is not fission to equal mass nuclei of about mass 120; the most common event (depending on isotope and process) is a slightly unequal fission in which one daughter nucleus has a mass of about 90 to 100 u and the other the remaining 130 to 140 u. Unequal fissions are energetically more favourable because this allows one product to be closer to the energetic minimum near mass 60 u (only a quarter of the average fissionable mass), while the other nucleus with mass 135 u is still not far out of the range of the most tightly bound nuclei (another statement of this, is that the atomic binding energy curve is slightly steeper to the left of mass 120 u than to the right of it).

The variation in specific binding energy with atomic number is due to the interplay of the two fundamental forces acting on the component nucleons (protons and neutrons) that make up the nucleus. Nuclei are bound by an attractive strong nuclear force between nucleons, which overcomes the electrostatic repulsion between protons. However, the strong nuclear force acts only over extremely short ranges, since it follows a Yukawa potential. For this reason large nuclei are less tightly bound per unit mass than small nuclei, and breaking a very large nucleus into two or more intermediate-sized nuclei releases energy.

Because of the short range of the strong binding force, large nuclei must contain proportionally more neutrons than do light elements, which are most stable with a 1-1 ratio of protons and neutrons. Extra neutrons stabilize heavy elements because they add to strong-force binding without adding to proton-proton repulsion. Fission products have, on average, about the same ratio of neutrons and protons as their parent nucleus, and are therefore usually unstable because they have proportionally too many neutrons compared to stable isotopes of similar mass. This is the fundamental cause of the problem of radioactive high level waste from nuclear reactors. Fission products tend to be beta emitters, emitting fast-moving electrons to conserve electric charge as excess neutrons convert to protons inside the nucleus of the fission product atoms.

The most common nuclear fuels, ^{235}U and ^{239}Pu, are not major radiologic hazards by themselves: ^{235}U has a half-life of approximately 700 million years, and although ^{239}Pu has a half-life of only about 24,000 years, it is a pure alpha particle emitter and hence not particularly dangerous unless ingested. Once a fuel element has been used, the remaining fuel material is intimately mixed with highly radioactive fission products that emit energetic beta particles and gamma rays. Some fission products have half-lives as short as seconds; others have half-lives of tens of thousands of years, requiring long-term storage in facilities such as Yucca mountain until the fission products decay into non-radioactive stable isotopes.

CHAIN REACTIONS

1. A uranium-235 atom absorbs a neutron and fissions into two new atoms (fission fragments), releasing three new neutrons and some binding energy.

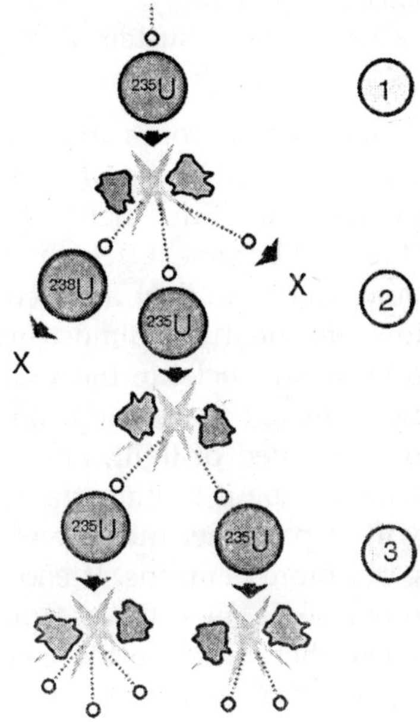

Fig. A schematic nuclear fission chain reaction.

2. One of those neutrons is absorbed by an atom of uranium-238 and does not continue the reaction. Another neutron is simply lost and does not collide with anything, also not continuing the reaction. However one neutron does collide with an atom of uranium-235, which then fissions and releases two neutrons and some binding energy.
3. Both of those neutrons collide with uranium-235 atoms, each of which fissions and releases between one and three neutrons, which can then continue the reaction.

Many heavy elements, such as uranium, thorium, and plutonium, undergo both spontaneous fission, a form of radioactive decay and *induced fission,* a form of nuclear reaction. Elemental isotopes that undergo induced fission when struck by a free neutron are called fissionable; isotopes that undergo fission when struck by a thermal, slow moving neutron are also called fissile. A few particularly fissile and readily obtainable isotopes (notably ^{235}U and ^{239}Pu) are called nuclear fuels because they can sustain a chain reaction and can be obtained in large enough quantities to be useful.

All fissionable and fissile isotopes undergo a small amount of spontaneous fission which releases a few free neutrons into any sample of nuclear fuel. Such neutrons would escape rapidly from the fuel and become a free neutron, with a half-life of about 15 minutes before they decayed to protons and beta particles. However, neutrons almost invariably impact and are absorbed by other nuclei in the vicinity long before this happens (newly-created fission neutrons are moving at about 7percent of the speed of light, and even moderated neutrons are moving at about 8 times the speed of sound). Some neutrons will impact fuel nuclei and induce further fissions, releasing yet more neutrons. If enough nuclear fuel is assembled into one place, or if the escaping neutrons are sufficiently contained, then these freshly generated neutrons outnumber the neutrons that escape from the assembly, and a *sustained nuclear chain reaction* will take place.

An assembly that supports a sustained nuclear chain reaction is called a critical assembly or, if the assembly is almost entirely made of a nuclear fuel, a critical mass. The word "critical" refers to a cusp in the behaviour of the differential equation that governs the number of free neutrons present in the fuel: if less than a critical mass is present, then the amount of neutrons is determined by radioactive decay, but if a critical mass or more is present, then the amount of neutrons is controlled instead by the physics of the chain reaction. The actual mass of a *critical mass* of nuclear fuel depends strongly on the geometry and surrounding materials.

Not all fissionable isotopes can sustain a chain reaction. ^{238}U, the most abundant form of uranium, is fissionable but not fissile: it undergoes induced fission when impacted by an energetic neutron with over 1 MeV of kinetic energy. But too few of the neutrons produced by ^{238}U fission are energetic enough to induce further fissions in ^{238}U, so no chain reaction is possible with this isotope. Instead, bombarding ^{238}U with slow neutrons causes it to absorb them (becoming ^{239}U) and decay by beta emission to ^{239}Np which then decays again by the same process to ^{239}Pu; that process is used to manufacture ^{239}Pu in breeder reactors, but does not contribute to a neutron chain reaction.

Fissionable, non-fissile isotopes can be used as fission energy source even without a chain reaction. Bombarding ^{238}U with fast neutrons induces fissions, releasing energy as long as the external neutron source is present. That effect is used to augment the energy released by modern thermonuclear weapons, by jacketing the weapon with ^{238}U to react with neutrons released by nuclear fusion at the centre of the device.

FISSION REACTORS

Critical fission reactors are the most common type of nuclear reactor. In a critical fission reactor, neutrons produced by fission of fuel atoms are used to induce yet more fissions, to sustain a controllable amount of energy release. Devices that produce engineered but non-self-sustaining fission reactions are *subcritical fission reactors*. Such devices use radioactive decay or particle accelerators to trigger fissions.

Critical fission reactors are built for three primary purposes, which typically involve different engineering trade-offs to take advantage of either the heat or the neutrons produced by the fission chain reaction:

- *Power reactors* are intended to produce heat for nuclear power, either as part of a generating station or a local power system such as a nuclear submarine.
- *Research reactors* are intended to produce neutrons and/or activate radioactive sources for scientific, medical, engineering, or other research purposes.
- *Breeder reactors* are intended to produce nuclear fuels in bulk from more abundant isotopes. The better known fast breeder reactor makes ^{239}Pu (a nuclear fuel) from the naturally very abundant ^{238}U (not a

nuclear fuel). Thermal breeder reactors previously tested using ^{232}Th continue to be studied and developed.

While, in principle, all fission reactors can act in all three capacities, in practice the tasks lead to conflicting engineering goals and most reactors have been built with only one of the above tasks in mind. Power reactors generally convert the kinetic energy of fission products into heat, which is used to heat a working fluid and drive a heat engine that generates mechanical or electrical power. The working fluid is usually water with a steam turbine, but some designs use other materials such as gaseous helium. Research reactors produce neutrons that are used in various ways, with the heat of fission being treated as an unavoidable waste product. Breeder reactors are a specialized form of research reactor, with the caveat that the sample being irradiated is usually the fuel itself, a mixture of ^{238}U and ^{235}U.

For a more detailed description of the physics and operating principles of critical fission reactors. For a description of their social, political, and environmental aspects.

FISSION BOMBS

One class of nuclear weapon, a *fission bomb* (not to be confused with the *fusion bomb*), otherwise known as an *atomic bomb* or *atom bomb,* is a fission reactor designed to liberate as much energy as possible as rapidly as possible, before the released energy causes the reactor to explode (and the chain reaction to stop). Development of nuclear weapons was the motivation behind early research into nuclear fission: the Manhattan Project of the U.S. military during World War II carried out most of the early scientific work on fission chain reactions, culminating in the Little Boy and Fat Man and Trinity bombs that were exploded over test sites, the cities Hiroshima, and Nagasaki, Japan in August of 1945.

Even the first fission bombs were thousands of times more explosive than a comparable mass of chemical explosive. Little Boy weighed a total of about four tons (of which 60 kg was nuclear fuel) and was 11 feet long; it also yielded an explosion equivalent to about 15,000 tons of TNT, destroying a large part of the city of Hiroshima. Modern nuclear weapons (which include a thermonuclear *fusion* as well as one or more fission stages) are literally hundreds of times more energetic for their weight than the first pure fission atomic bombs, so that a modern single missile warhead bomb weighing less than 1/8th as much as Little Boy has a yield of 475,000 tons of TNT, and could bring destruction to 10 times the city area.

While the fundamental physics of the fission chain reaction in a nuclear weapon is similar to the physics of a controlled nuclear reactor, the two types of device must be engineered quite differently. It is impossible to convert a nuclear reactor to cause a true nuclear explosion, or for a nuclear reactor to explode the way a nuclear explosive does, (though partial fuel meltdowns and steam explosions have occurred), and similarly difficult to extract useful power from a nuclear explosive (though at least one rocket propulsion system, Project Orion, was intended to work by exploding fission bombs behind a massively padded vehicle!).

The strategic importance of nuclear weapons is a major reason why the technology of nuclear fission is politically sensitive. Viable fission bomb designs are, arguably, within the capabilities of bright undergraduates (being incredibly simple, but nuclear fuel to realise the designs is thought to be difficult to obtain being rare.

While many believe that Ernest Rutherford became the first person to deliberately split the atom by bombarding nitrogen with naturally occurring alpha particles from radioactive material and observing a proton emitted with energy higher than the alpha particle, this is incorrect. Rutherford forced protons out of the nucleus but did not split the nucleus, as a fission reaction requires. His students1932 John Cockcroft and Ernest Walton, working under Rutherford's direction, attempted to split the nucleus by entirely artificial means, using a particle accelerator to bombard lithium with protons thereby producing two alpha particles. This still did not result in nuclear fission.

Results of the bombardment of uranium by neutrons had proved interesting and puzzling. First studied by Enrico Fermi and his colleagues in 1934, they were not properly interpreted until several years later.

After the Fermi publication, Lise Meitner, Otto Hahn and Fritz Strassmann began performing similar experiments in Germany. Meitner, an Austrian Jew, lost her citizenship with the Anschluss in 1938. She fled and wound up in Sweden, but continued to collaborate by mail and through meetings with Hahn in Sweden. By coincidence her nephew Otto Robert Frisch, also a refugee, was also in Sweden when Meitner received a letter from Hahn describing his chemical proof that some of the product of the bombardment of uranium with neutrons was barium (barium's atomic weight is half that of uranium). Frisch was skeptical, but Meitner believed Hahn was too good a chemist to have made a mistake. According to Frisch:

No larger fragments than protons or helium nuclei (alpha particles) had ever been chipped away from nuclei, and to chip off a large number not nearly enough energy was available. Nor was it possible that the uranium nucleus could have been cleaved right across. A nucleus was not like a brittle solid that can be cleaved or broken; George Gamow had suggested early on, and Bohr had given good arguments that a nucleus was much more like a liquid drop. Perhaps a drop could divide itself into two smaller drops in a more gradual manner, by first becoming elongated, then constricted, and finally being torn rather than broken in two? We knew that there were strong forces that would resist such a process, just as the surface tension of an ordinary liquid drop tends to resist its division into two smaller ones. But nuclei differed from ordinary drops in one important way: they were electrically charged, and that was known to counteract the surface tension.

The charge of a uranium nucleus, we found, was indeed large enough to overcome the effect of the surface tension almost completely; so the uranium nucleus might indeed resemble a very wobble unstable drop, ready to divide itself at the slightest provocation, such as the impact of a single neutron. But there was another problem. After separation, the two drops would be driven apart by their mutual electric repulsion and would acquire high speed and hence a very large energy, about 200 MeV in all; where could that energy come from?...Lise Meitner... worked out that the two nuclei formed by the division of a uranium nucleus together would be lighter than the original uranium nucleus by about one-fifth the mass of a proton. Now whenever mass disappears energy is created, according to Einstein's formula $E=mc^2$, and one-fifth of a proton mass was just equivalent to 200MeV. So here was the source for that energy; it all fitted!

The basic discovery and chemical proof of Otto Hahn and Fritz Strassmann that an isotope of barium was produced by neutron bombardment of uranium was published in a paper in Germany in the Journal *(Naturwissenschaften,* January 6, 1939) and earned Hahn a Nobel Prize.

Frisch rapidly confirmed experimentally by means of a cloud chamber that the uranium atom had indeed been split by the action of neutrons. A fundamental idea of this experiment was suggested to Frisch by George Placzek. Two papers were mailed to England on January 16, 1939, the first on the interpretation of the barium appearance as atom splitting by Meitner and Frisch, the second on the experimental confirmation by Frisch (strangely omitting Placzek's important contribution, however). The first paper appeared on February 11, the second on February 28.

Meitner and Frisch's theory and mathematical proof of Hahn's discovery and chemical proof of barium products from the bombardment of uranium was the foundation of the later research on nuclear fission. The awarding of the 1944 Nobel Prize in Chemistry to Hahn alone is a longstanding controversy.

On January 16, 1939, Niels Bohr of Copenhagen, Denmark, arrived in the United States to spend several months in

Princeton, New Jersey, and was particularly anxious to discuss some abstract problems with Albert Einstein. Just before Bohr left Denmark, Frisch and Meitner gave him their calculations.

Bohr had promised to keep the Meitner/Frisch paper secret until it was published to preserve priority, but on the boat he discussed it with Léon Rosenfeld, and forgot to tell him to keep it secret. Rosenfeld immediately upon arrival told everyone at Princeton University, and from them the news spread by word of mouth to neighbouring physicists including Enrico Fermi at Columbia University. Fermi upon traveling to receive the Nobel Prize for his earlier work. headed to the USA rather than return to Fascist Italy with his Jewish wife. As a result of conversations among Fermi, John R. Dunning, and G. B. Pegram, a search was undertaken at Columbia for the heavy pulses of ionization that would be expected from the flying fragments of the uranium nucleus. On January 26, 1939, there was a conference on theoretical physics at Washington, D.C., sponsored jointly by the George Washington University and the Carnegie Institution of Washington. Before the meeting in Washington was over, several other experiments to confirm fission had been initiated, and positive experimental confirmation was reported.

Frédéric Joliot-Curie's team in Paris discovered that secondary neutrons are released during uranium fission thus making a chain reaction feasible. About two neutrons being emitted with nuclear fission of uranium was verified independently by Leo Szilard and Walter Zinn. The number of neutrons emitted with nuclear fission of 235uranium was then reported at 3.5/fission, and later corrected to 2.6/fission by Frédéric Joliot-Curie, Hans von Halban and Lew Kowarski.

"Chain reactions" at that time were a known phenomenon in *chemistry*, but the analogous process in nuclear physics using neutrons had been foreseen as early as 1933 by Leo Szilard, although Szilard at that time had no idea with what materials the process might be initiated. Szilard, a Hungarian born Jew, also fled mainland Europe after Hitler's rise, eventually landing in the US.

In the summer Fermi and Szilard proposed the idea of a nuclear reactor (pile) with natural uranium as fuel and graphite as moderator of neutron energy.

In August Hungarian-Jewish refugees Szilard, Teller and Wigner persuaded Austrian-Jewish refugee Einstein to warn President Roosevelt of the German menace. The letter suggested the possibility of uranium bomb deliverable by ship. The President received it on 1939.10.11 shortly after WWII began.

In England James Chadwick proposed an atomic bomb utilizing natural uranium based on a paper by Rudolf Peierls with the mass needed for critical state being 30-40 tons.

In December, Heisenberg delivered a report to the Germany Department of War on the possibility of a uranium bomb.

In Birmingham, England Otto Robert Frisch teamed up with Rudolf Peierls who had also fled German anti-Jewish race laws. They conceived the idea of utilizing a purified isotope of uranium, uranium-235, and worked out that an enriched uranium bomb could have a critical mass of only 600 g. instead of tons, and that the resulting explosion would be tremendous. (the amount actually turned out to be 15 kg.) In February 1940 they delivered the Frisch-Peierls memorandum, however, they were officially considered "enemy aliens" at the time.

Glenn Seaborg, Joe Kennedy, Art Wahl and Italian-Jewish refugee Emilio Segre discovered plutonium and determined it to be fissionable like U-235. (Lawrence controversially dropped Segre's pay by half when he learned he was trapped in the US by Mussolini's race laws.)

On June 28 1941, the Office of Scientific Research and Development was formed to mobilize scientific resources and apply the results of research to national defence. In September Fermi assembled his first nuclear pile in an attempt to create a slow neutron induced chain reaction in uranium. but the experiment failed.

Producing a fission chain reaction in uranium fuel is far from trivial. Early nuclear reactors did not use isotopically enriched uranium, and in consequence they were required to use large quantities of highly purified graphite as neutron moderation materials. Use of ordinary water (as opposed to heavy water) in nuclear reactors requires enriched fuel— the partial separation and relative enrichment of the rare ^{235}U isotope from the far more common ^{238}U isotope. Typically, reactors also require inclusion of extremely chemically pure neutron moderator materials such as deuterium (in heavy water), helium, beryllium, or carbon, usually as the graphite. (The high purity is required because many chemical impurities such as the boron-10 component of natural boron, are very strong neutron absorbers and thus poison the chain reaction.)

Production of such materials at industrial scale had to be solved for nuclear power generation and weapons production to be accomplished. Up to 1940, the total amount of uranium metal produced in the USA was not more than a few grams and even this was of doubtful purity; of metallic beryllium not more than a few kilograms; concentrated deuterium oxide (heavy water) not more than a few kilograms; and finally carbon had never been produced in quantity with anything like the purity required of a moderator.

The problem of producing large amounts of high purity uranium was solved by Frank Spedding using the thermite process. Ames Laboratory was established in 1942 to produce the large amounts of natural (unenriched) uranium that would be necessary for the research to come. The success of the Chicago Pile-1 which used unenriched (natural) uranium, like all of the atomic "piles" which produced the plutonium for the atomic bomb, was also due specifically to Szilard's realization that very pure graphite could be used for the moderator of even natural uranium "piles". In wartime Germany, failure to appreciate the qualities of very pure graphite led to reactor designs dependent on heavy water, which in turn was denied the Germans by Allied attacks in

Norway, where heavy water was produced. These difficulties prevented the Nazis from building a nuclear reactor capable of criticality during the war.

Unknown until 1972, when French physicist Francis Perrin discovered the Oklo Fossil Reactors, nature had beaten humans to the punch by engaging in large-scale uranium fission chain reactions, some 2,000 million years in the past. This ancient process was able to use normal water as a moderator, only because 2,000 million years in the past, natural uranium was "enriched" with the shorter-lived fissile isotope ^{235}U, as compared with the natural uranium available today.

Natural Nuclear Fission Reactor

Geological Situation in Gabon leading to natural nuclear fission reactors

1. Nuclear reactor zones
2. Sandstone
3. Uranium ore layer
4. Granite

Fig. Geological Situation in Gabon

A natural nuclear fission reactor is a uranium deposit where analysis of isotope ratios has shown that self-sustaining nuclear chain reactions have occurred. The existence of this phenomenon was discovered in 1972 by French physicist Francis Perrin. The conditions under which a natural nuclear reactor could exist were predicted at the University of Arkansas by Paul Kuroda in 1956. The conditions found at Oklo were very similar to what Kuroda predicted.

At the only known location, three ore deposits at Oklo in Gabon, sixteen sites have been discovered so far at which self-sustaining nuclear fission reactions took place approximately 1.5 billion years ago, and ran for a few hundred thousand years, averaging 100 kW of power output during that time.

In May 1972 at the Pierrelatte uranium enrichment facility in France, routine mass spectrometry comparing UF_6 samples from the Oklo Mine, located in Gabon, Central Africa, showed a discrepancy in the amount of the ^{235}U isotope. Normally the concentration is 0.7202percent; these samples had only 0.7171percent – a significant difference. This discrepancy required explanation, as all uranium handling facilities must meticulously account for all fissionable isotopes to assure that none are diverted for weapons purposes. Thus the French Commissariat à l'énergie atomique began an investigation. A series of measurements of the relative abundances of the two most significant isotopes of the uranium mined at Oklo showed anomalous results compared to those obtained for uranium from other mines. Further investigations into this uranium deposit discovered uranium ore with a ^{235}U to ^{238}U ratio as low as 0.440percent.

Fission Product Isotope Signatures

Nd

Neodymium and other elements were found with isotopic compositions different from what is customarily found on Earth. Natural neodymium contains 27percent ^{142}Nd; the Nd at Oklo contained less than 6percent but contained more ^{143}Nd.

Subtracting the natural isotopic Nd abundance from the Oklo-Nd, the isotopic composition matched that produced by the fissioning of ^{235}U.

A diagram showing the isotope signatures of natural neodymium and fission product neodymium from U-235 which had been subjected to thermal neutrons. Note that the

Ce-142 (a long lived beta emitter) has not had time to decay to Nd-142 over the time since the reactors stopped working.

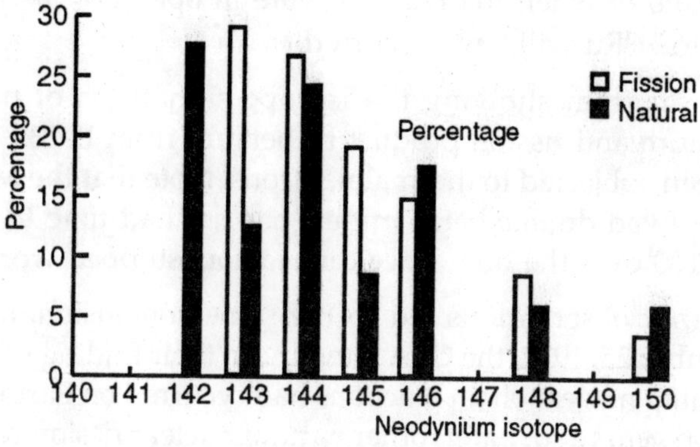

Ru

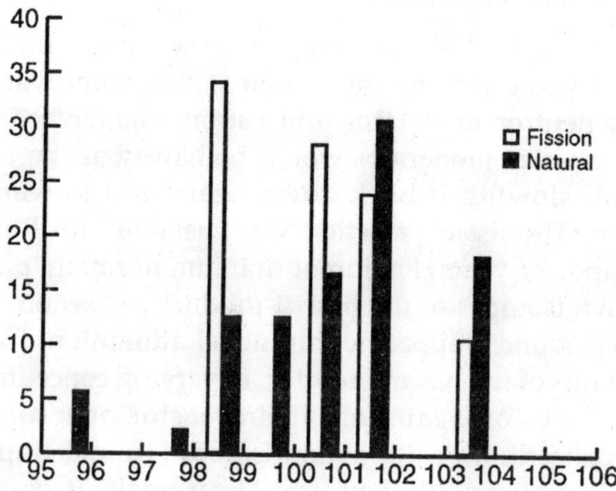

Similar investigations into the isotopic ratios of ruthenium at Oklo found a much higher ^{99}Ru concentration than expected (27-30percent vs. 12.7percent). This anomaly could be explained by the decay of ^{99}Tc to ^{99}Ru. In the bar chart below the normal natural isotope signature of ruthenium is compared with that for fission product ruthenium which is the result of the fission of ^{235}U with thermal neutrons. It is clear that the fission ruthenium has a different isotope signature. The level of ^{100}Ru in the fission product mixture is low because of a long lived (half life = 10^{19} years) isotope of molybdenum. On the time scale of when the reactors were in operation very little decay to ^{100}Ru will have occurred.

A diagram showing the isotope signatures of natural ruthenium and fission product ruthenium from U-235 which had been subjected to thermal neutrons. Note that the Mo-100 (a long lived double beta emitter) has not had time to decay to Ru-100 over the time since the reactors stopped working.

Other observations led to the same conclusion, and on September 25, 1972, the CEA announced their finding that self-sustaining nuclear chain reactions had occurred on Earth about 2 billion years ago. Later, other natural nuclear fission reactors were discovered in the region.

Mechanism of The Reactors

The natural nuclear reactor formed when a uranium-rich mineral deposit became inundated with groundwater that acted as a neutron moderator, and a strong chain reaction took place. The water moderator would boil away as the reaction increased, slowing it back down again and preventing a meltdown. The fission reaction was sustained for hundreds of thousands of years. Fission of uranium normally produces five known isotopes of the fission-product gas xenon; all five have been found trapped within novel aluminium foams in the remnants of the natural reactor, in varying concentrations. This points to on-again, off-again reactor operation. The specific concentrations of xenon isotopes, found trapped in mineral formations 2 billion years later, make it possible to calculate the specific time intervals of reactor operation: approximately 2 hours and 30 minutes.

It is estimated that secondary enrichment of the uranium in centimeter- to meter-sized veins consumed about six tons of ^{235}U and elevated temperatures to a few hundred degrees Celsius. Remarkably, the non-volatile fission products have only moved a few centimeters in the veins during the last 1.5 billion years. This offers a case study of how radioactive isotopes migrate through the earth's crust—a significant area of controversy as opponents of geologic nuclear waste disposal fear that releases from stored waste could end up in water supplies or be carried into the environment.

A key factor that made the reaction possible was that, at the time the reactor went critical, the fissile isotope ^{235}U made up about 3percent of the natural uranium, which is comparable to the amount used in some of today's reactors. (The remaining 97percent was non-fissile ^{238}U.) Because ^{235}U has a shorter half life than ^{238}U, and thus decays more rapidly, the current abundance of ^{235}U in natural uranium is about 0.7percent. A natural nuclear reactor is therefore no longer possible on Earth.

Another factor which probably contributed to the operation of the Oklo natural nuclear reactor was the fact that uranium is soluble in water only in the presence of oxygen. Increasing oxygen content in the earth's atmosphere helps explain why this natural reactor started operation sometime around 2 billion years ago and not before, since the percentage abundance of fissionable ^{235}U was at least 3percent or higher at all times before reactor startup. Rising oxygen levels during the aging of earth may have allowed a high enough concentration of uranium to accumulate to enhance the possibility of reactor startup.

When oxygenated water trickled over uranium deposits and picked up a few ppm of uranium, uranium may have been re-deposited and concentrated due to the action of algae and geological placer formations more than once. Other geological action, such as uplifting and fracturing of the concentrated uranium ore formations, may have caused further concentrations into a compacted vein. Without the new aerobic environment available on earth at the time, these concentrations probably couldn't have taken place.

The natural reactor of Oklo can also be used to check if the fine-structure constant á might have changed over time. Alex Shlyakhter proposed in 1976 to measure the abundance of ^{149}Sm to estimate the cross section for neutron capture of this isotope at the time of the reactions and check it against the present value. Measurements show that the constant has not changed

Relation to Yucca Mountain and other Geologic Repositories

The deep geological repository concept involves the encapsulation of used fuel in long-lived engineered containers which are then placed and sealed within excavated rooms in a naturally occurring geological formation at a design depth of 500 to 1000 metres below ground surface.

The ability of natural geologic barriers to isolate radioactive waste is demonstrated by the Oklo reactors. During their long reaction period about 5.4 tonnes of fission products as well as 1.5 tonnes of plutonium together with other transuranic elements were generated in the orebody. This plutonium and the other transuranics remained immobile until the present day. This is quite remarkable because ground water had ready access to the deposits and they were not in a chemically inert form, such as glass.

Thus the only known example of underground nuclear waste disposal was successful over a long period in spite of the characteristics of the site. Such a water-logged, sandstone/shale structure would not be considered for disposal of modern toxic wastes, nuclear or otherwise, although the clays and bitumen present played an important part in containing the material.

BASIC NUCLEAR FISSION

Nuclear fission is the process of splitting atoms, or fissioning them. This topic will explain the basics of nuclear fission. Everyone's played with marbles at one time or another, right? Well, imagine about 200 marbles lying on a flat surface, all jumbled together, and roughly forming a circle. What would happen if someone took another marble and threw it at them? They would fly all around in different directions and groups, right? That is exactly what happens in nuclear fission. The filled circle is like an atom's nucleus. The marble being thrown is like a "neutron bullet". The only differences are that the marbles are protons and neutrons and the protons and neutrons aren't in a filled circle, but in the actual atom are in the shape of a sphere. Of course, an atom is also a bit more complicated than a pack of marbles.

Choosing the Bullet

When we spoke about the marble analogy earlier, we said that the marble being thrown is a like a "neutron bullet". But what does this mean, and why not use another type of particle to "throw" at a nucleus to fission it? First, what particles with distinct mass are available to launch at a nucleus? Think back to our lesson on radioactivity. Recall that two particles emitted by radioactive elements are the α particle and the neutron. (There are other particles emitted too, but they are generally much smaller than the neutron and the α particle.) Recall that the α particle is essentially a ^{4}He nucleus. Now, let's review the structure of an atom. Remember that an atomic nucleus is made up of positive protons and neutral neutrons? Because of this, the nucleus carries an overall positive charge. So, if we were to launch another particle with a positive charge at a nucleus, it wouldn't get there. Why wouldn't it get there? The answer lies in magnetism. Have you ever used magnets? If you have, you'd know that two like poles of a magnet repel each other. A positive particle and the positive nucleus would repel each other in the same way. The α particle is positive. Why? Well, it's composed of two protons and two neutrons. Its positive protons give it a positive charge. Because it's positive, it would get repelled away from another positive nucleus. So, the only thing left is the neutron. The neutron is electrically neutral and thus would not get repelled from a positive nucleus.

Fissile Isotopes

Fissile isotopes are isotopes of an element that can be split through fission. Only certain isotopes of certain elements are fissile. One isotope of uranium, $_{235}$U, is fissile, while another isotope, $_{238}$U, is not. Other examples of fissile elements are

$_{239}$Pu and $_{232}$Th. An important factor affecting whether or not an atom will fission is the speed at which the bombarding neutron is moving. If the neutron is highly energetic (and thus moving very quickly), it can cause fission in some elements that a slower neutron would not. Thorium 232 requires a very fast neutron to induce fission. However, uranium 235 needs slower neutrons. If a neutron is too fast, it will pass right through a $_{235}$U atom without affecting it at all.

Splitting the Uranium Atom

Uranium is the principle element used in nuclear reactors and in certain types of atomic bombs. The specific isotope used is $_{235}$U. When a stray neutron strikes a $_{235}$U nucleus, it is at first absorbed into it. This creates $_{236}$U. $_{236}$U is unstable and this causes the atom to fission. The fissioning of $_{236}$U can produce over twenty different products. However, the products' masses always add up to 236. The following two equations are examples of the different products that can be produced when $_{235}$U fissions:

$$_{235}\text{U} + 1 \text{ neutron} \rightarrow 2 \text{ neutrons} + {}_{92}\text{Kr} + {}_{142}\text{Ba} + \text{ENERGY}$$

$$_{235}\text{U} + 1 \text{ neutron} \rightarrow 2 \text{ neutrons} + {}_{92}\text{Sr} + {}_{140}\text{Xe} + \text{ENERGY}$$

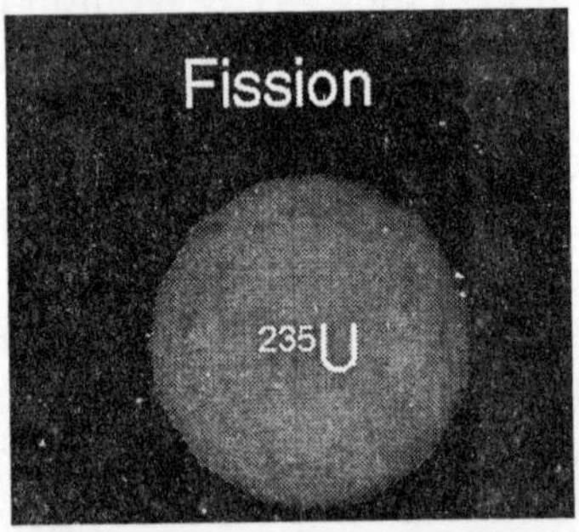

Fig. Fission of Uranium Atom

Let's discuss those reactions. In each of the reactions, 1 neutron splits the atom. When the atom is split, 1 additional neutron is released. This is how a chain reaction works. If more $_{235}$U is present, those 2 neutrons can cause 2 more atoms to split. Each of those atoms releases 1 more neutron bringing the total neutrons to 4. Those **4 neu**trons can strike 4 more $_{235}$U atoms, releasing even more neutrons. The chain reaction will continue until all the $_{235}U$ fuel is spent. This is roughly what happens in an atomic bomb. It is called a runaway nuclear reaction.

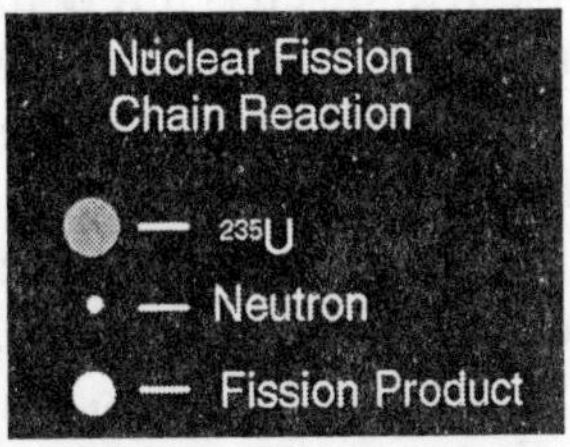

Fig. Fission of Uranium Atom

The fissioning of each 235U atom (red) releases more neutrons (green) that go on to fission more $_{235}U$ atoms, thus producing a chain reaction.

Where Does the Energy Come From?

In the section above we described what happens when an $_{235}U$ atom fissions. We gave the following equation as an example:

$$_{235}U + 1 \text{ neutron} \rightarrow 2 \text{ neutrons} + {}_{92}Kr + {}_{142}Ba + \text{ENERGY}$$

It might have been a amazing question, "Where does the energy come from?". The mass seems to be the same on both sides of the reaction:

$$235 + 1 = 2 + 92 + 142 = 236$$

Thus, it seems that no mass is converted into energy. However, this is not entirely correct. The mass of an atom is more than the sum of the individual masses of its protons and neutrons, which is what those numbers represent. Extra mass is a result of the binding energy that holds the protons and neutrons of the nucleus together. Thus, when the uranium atom is split, some of the energy that held it together is released as radiation in the form of heat. Because energy and mass are one and the same, the energy released is also mass released. Therefore, the total mass does decrease a tiny bit during the reaction.

THE NUCLEAR FISSION POWER PLANT

Currently, about half of all nuclear power plants are located in the US. There are many different kinds of nuclear power plants, and we will discuss a few important designs in this text. A nuclear power plant harnesses the energy inside atoms themselves and converts this to electricity. This electricity is used by all of us. By now, you should have an idea of the fission process and how it works. A nuclear power plant uses controlled nuclear fission. We will explore how a nuclear power plant operates and the manner in which nuclear reactions are controlled.

Uranium Preparation

Earlier we talked about nuclear fission with $_{235}U$. In reality, this will not be the only isotope of uranium present in a nuclear reactor. In naturally occurring uranium deposits, less than one percent of the uranium is $_{235}U$. The majority of the uranium is $_{238}U$. $_{238}U$ is not a fissile isotope of uranium. When $_{238}U$ is struck by a loose neutron, it absorbs the neutron into its nucleus and does not fission. Thus, by absorbing loose neutrons, $_{238}U$ can prevent a nuclear chain reaction from occurring. This would be a bad thing because if a chain reaction doesn't occur, the nuclear reactions can't sustain themselves, the reactor shuts down, and millions of people are without electrical power. In order for a chain reaction to occur, the pure uranium ore must be refined to raise the concentration of $_{235}U$. This is called enrichment and is primarily accomplished through a technique called gaseous diffusion. In this process, the uranium ore is combined with fluorine to create a chemical compound called uranium hexafluoride. The uranium hexafluoride is heated and vapourizes. The heated gas is then pushed through a series of filters. Because some of the uranium hexafluoride contains 238U and some contains $_{235}U$, there is a slight difference in the weights of the individual molecules. The molecules of uranium hexafluoride containing $_{235}U$ are slightly lighter and thus pass more easily through the filters. This creates a quantity of uranium hexafluoride with a higher proportion of $_{235}U$. This is collected, the uranium is stripped from it, and the result is an enriched supply of fuel. Usually, nuclear power plants use uranium fuel that is about 4percent $_{235}U$.

Parts of a Nuclear Reactor - Pressurized Water Reactor (PWR):

A typical nuclear reactor has a few main parts. Inside the "core" where the nuclear reactions take place are the fuel rods and assemblies, the control rods, the moderator, and the coolant. Outside the core are the turbines, the heat exchanger, and part of the cooling system.

The fuel assemblies are collections of fuel rods. These rods are each about 3.5 meters (11.48 feet) long. They are each about a centimeter in diameter. These are grouped into large bundles of a couple hundred rods called fuel assemblies, which are then placed in the reactor core. Inside each fuel rod are hundreds of pellets of uranium fuel stacked end to end.

Also in the core are control rods. These rods have pellets inside that are made of very efficient neutron capturers. An example of such a material is cadmium. These control rods are connected to machines that can raise or lower them in the core. When they are fully lowered into the core, fission can not occur because they absorb free neutrons. However, when they are pulled out of the reactor, fission can start again anytime a stray neutron strikes a $_{235}U$ atom, thus releasing more neutrons, and starting a chain reaction.

Another component of the reactor is the moderator. The moderator serves to slow down the high speed neutrons "flying" all around the reactor core. If a neutron is moving too fast, and thus is at a high-energy state, it passes right through the $_{235}U$ nucleus. It must be slowed down to be captured by the nucleus and to induce fission. The most common moderator is water, but sometimes it can be another material.

The job of the coolant is to absorb the heat from the reaction. The most common coolant used in nuclear power plants today is water. In actuality, in many reactor designs the coolant and the moderator are one and the same. The coolant water is heated by the nuclear reactions going on inside the core. However, this heated water does not boil because it is kept at an extremely intense pressure, thus raising its boiling point above the normal 100° Celsius.

The heated water rises up and passes through another part of the reactor, the heat exchanger. The moderator/coolant water is radioactive, so it can not leave the inner reactor containment. Its heat must be transferred to non-radioactive water, which can then be sent out of the reactor shielding. This is done through the heat exchanger, which works by moving the radioactive water through a series of pipes that are wrapped around other pipes. The metallic pipes conduct the heat from the moderator to the normal water. Then, the normal water (now in steam form and intensely hot) moves to the turbine, where electricity is produced.

After the hot water has passed through the turbine, some of its energy is changed into electricity. However, the water is still very hot. It must be cooled somehow. Many nuclear power plants used steam towers to cool this water with air. These are generally the buildings that people associate with nuclear power plants. At reactors that do not have towers, the clean water is purified and dumped into the nearest body of water, and cool water is pumped in to replace it.

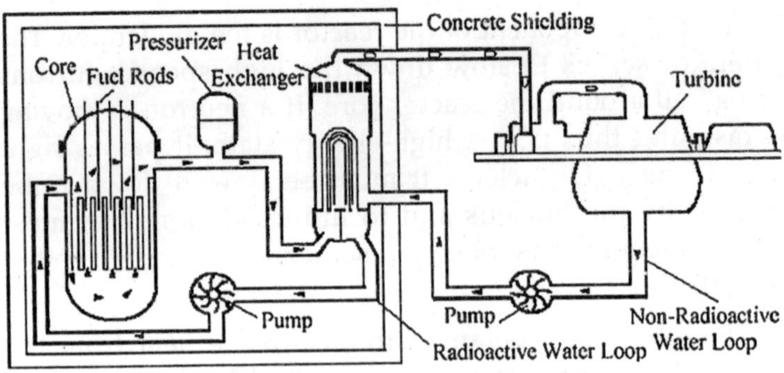

Fig. PWR Power Plant Schematic

From Fission to Electricity

A nuclear power plant produces electricity in almost exactly the same way that a conventional (fossil fuel) power plant does. A conventional power plant burns fuel to create heat. The fuel is generally coal, but oil is also sometimes used. The heat is used to raise the temperature of water, thus causing it to boil. The high temperature and intense pressure steam that results from the boiling of the water turns a turbine, which then generates electricity. A nuclear power plant works the same way, except that the heat used to boil the water is produced by a nuclear fission reaction using $_{235}U$ as fuel, not the combustion of fossil fuels. A nuclear power plant uses much less fuel than a comparable fossil fuel plant. A rough estimate is that it takes 17,000 kilograms of coal to produce the same amount of electricity as 1 kilogram of nuclear uranium fuel.

Other Types of Reactors

Although the most common type of reactor is the Pressurized Water Reactor (PWR), many other types of reactors are also used. In the PWR, as we described earlier, there are two main water cycles. One is the water inside the core that is highly radioactive. This water's heat is transferred to other, non-radioactive water inside the second loop. This water is then used to turn a turbine. The second most popular reactor type is the Boiling Water Reactor (BRW). This type of reactor differs from the PWR in that there is only one water cycle. Radioactive water is used to turn the turbine. The major disadvantage of this is that the radioactive nuclides in the water that cause its radioactivity can be transferred to the turbine, thus causing it to become radioactive too. This produces more hazardous material that needs to be disposed of when a reactor is dismantled. However, the BWR also has a few advantages. Its core can be kept at a lower pressure.

Another type of reactor is the Heavy Water Reactor (HWR). A HWR uses heavy water as a moderator instead of normal water. Heavy water is water with deuterium, which is an isotope of hydrogen with 1 neutron. Deuterium is heavier than normal hydrogen, which has no neutrons. HWR's come in two types, pressurized and boiling, just like normal "light water" reactors. The advantage of a HWR is that un-enriched uranium fuel can be used. This is because the heavy water is a much more efficient moderator than light water. Thus, more stray neutrons can be slowed down enough to cause fission in 235U. This more efficient moderator makes up for the greater abundance of the neutron-capturing $_{238}$U.

NUCLEAR CHAIN REACTION

In a fission reaction, besides the two fission products, a few (in the case of U-235 2.47 on the average) neutrons are emitted. The principle of chain reaction is very easy therefore: a U235 nucleus is hit by a neutron, upon which fission occurs at a high probability. The neutrons produced in this process can cause more fissions, moreover, the already mentioned 200 MeV energy is released each time. Consequently, the process becomes self-sustaining and no external neutron source is needed any longer.

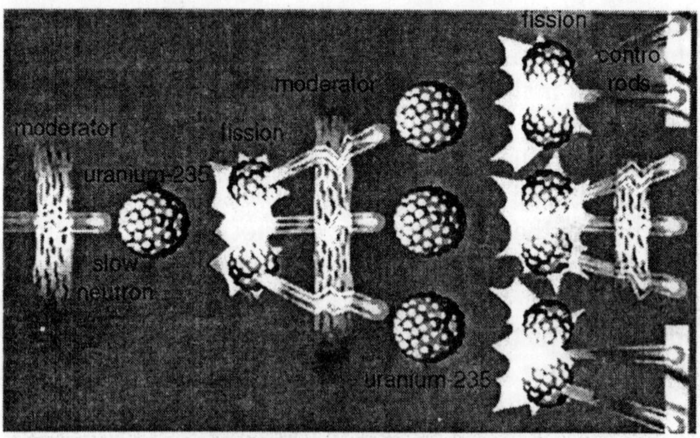

Fig. Nuclear chain reaction

If the neutrons produced from a fission induce only one new fission (because of escaping from the system or being captured inside) we say that the chain reaction is of constant intensity. If the neutrons generated from a fission reaction induce more than one fission than more and more fission reactions will take place in a unit time and the chain reaction will be diverging.

The probability that a high energy neutron emitted in a fission reaction can directly cause fission is quite low. In order to build a system in which fast neutrons sustain the chain reaction, very highly enriched uranium is needed, which is very expensive. A more easily feasible way is to use materials that slow the neutrons down to such low energies at which the probability of causing a fission is significantly higher. These neutron slowing down materials are the so called moderators.With the aid of some adequate moderator one might as well achieve chain reaction using natural (0.7percent U235 content) uranium.

If we simplify things a little, there are two major requirements that a moderator material should meet: it should have as low atomic mass number as possible and its neutron absorbing ability should be as low as possible. In practice four elements meet these requirements: water (H_2O), heavy-water (D_2O), graphite (C) and beryllium (Be). Among these water is the most widespread moderator. The properties of heavy-water are actually better but its disadvantage is the very high price.

If we start to put together more and more uranium (of course mixed with some moderator), the expected chain reaction will not start for a good while. The reason for this is that in the case of a small amount of fissile material large part of the neutrons that originate in fission will leak out of the pile and therefore they will not be able to cause fission. By increasing the amount of fissile material there will be a moment when we reach the so called *critical mass* and a self sustaining reaction starts. For pure metal U235 (without moderator) the critical mass is somewhat less than 50 kg. Because of the very high density (19.2 g/cm^3) this amount of uranium makes a sphere whose diameter is about 17 cm.

As we have already mentioned, at the fission of U-235 on the average 2.5 neutrons are released but not all of these cause fission. The neutronic state can be characterized by the so called *multiplication factor (k)*. This is the ratio of the number of neutrons in a neutron "generation" and the number of neutons in the preceding generation. (Neutron generation is the time that elapses between the birth of a neutron and the birth of neutrons from the subsequent fission which is caused by the given neutron. Its tipical value is about 10^{-9} s [i.e. 1 nanosecond] for pure U-235.)

If the value of the multiplication factor is 1, we talk about a *critical* reactor. In this case the number of neutrons in the system is constant, i.e. they cause the same number of fissions in every second. If $k<1$, the reactor is *subcritical* and the number of neutrons is going down. On the other hand, if $k>1$, the system is *supercritical*.

Very often the notion of *reactivity* is used instead of the multiplication factor. The definition of reactivity is $\rho=(k-1)/k$.

Correspondingly, for a subcritical reactor $\rho<0$, for a supercritical $\rho>0$, and for a critical $\rho=0$.

Obviously, the number of neutrons present in the reactor must be regulated or controlled since this determines the rate of fission and thus the energy released per second. In order to *control* the chain reaction one should use materials which tend to capture neutrons at a high probability. The most widespread neutron absorbers are cadmium (Cd) and boron (B).

The so called *control rods* are important tools of control and can be found in every reactor. These are made of neutron absorbing materials and can be moved between the fuel assemblies. If one wishes to decrease the power of the reactor it is sufficient to push a control rod a little inner. The control rods are particularly useful for the short term control and stopping the reactor. For long term regulation usually boric acid dissolved in the coolant is used.

The majority of the emitted neutrons leave the place of fission within an extremely short time. However, about 0.64percent of them (in the case of U-235) only leave the fission product significantly later. These are the so called delayed neutrons, which play a vital role in the controllability of reactors.

Change in the power results in changes in temperatures (fuel, coolant etc.). The temperature changes in turn modify the reactivity through different nuclear physics processes, which has a well defined influence on the number of neutrons ready to induce fission and thus on power. When the power rises, the temperature of fuel will rise too. However, the higher the temperature, the higher the ability of U-238 to absorb neutrons. Therefore, due to the presence of more neutron absorbing material reactivity is lower and the number of neutrons and accordingly power will decrease. Finally we may say that a chnge in power caused a change in reactivity and this in turn had an effect on power. If the change in reactivity is such that the processes caused are against the originating ones, we talk about negative feedback. The feedback was negative: the power increase induced a negative change in reactivity.

In terms of reactor safety it is an obvious intention that all feedbacks should be negative. In this case any power increase would cause a drop in reactivity and this will make the reactor sub-critical sooner or later, without external interaction. In such cases we say that the reactor has a so called inherent safety. Such inherent safety is characteristic for all PWRs and among them the reactors of the Paks NPP.

As a counter-example for inherent safety we may mention the RBMK type reactors, because in these positive feedbaks can occur in certain operational states. This contributed largely, among other factors, to the accident of the Chernobyl power plant.

Advantages

- Relatively little fuel is needed and the fuel is relatively inexpensive and available in trace amounts around the world.

- Fission is not believed to contribute to global warming or other pollution effects associated with fossil fuel combustion

Disadvantages

- Possibility of nuclear meltdown from uncontrolled reaction—leads to nuclear fallout with potentially harmful effects on civilians
- Waste products can be used to manufacture weapons

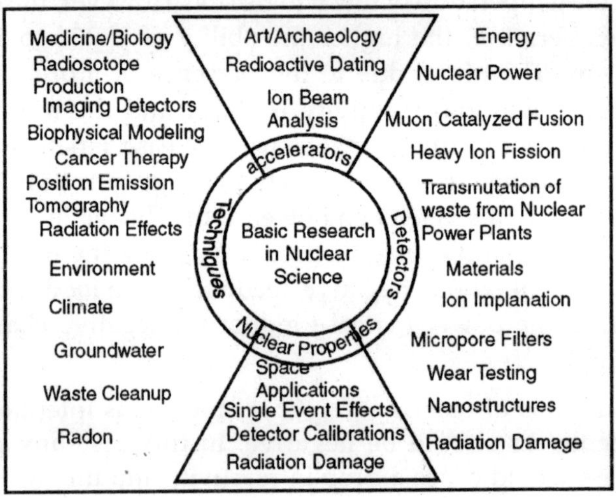

- High initial cost because plant requires containment safeguards

Smoke Detectors

Most common smoke detectors contain a small amount of ^{241}Am, a radioactive isotope. ^{241}Am is produced and recovered from nuclear reactors. Alpha particles emitted by the decays of ^{241}Am ionize the air (split the air molecules into electrons and positive ions) and generate a small current of electricity that is measured by a current-sensitive circuit. When smoke enters the detector, ions become attached to the smoke particles, which causes a decrease in the detector current. When this happens, an alarm sounds. These detectors provide warning for people to leave burning homes safely. Many lives have been saved by the their use.

Because the distance alpha particles travel in air is so short, there is no risk of being exposed to radiation by having a smoke detector in the house. Since ionization-type smoke detectors contain radioactive materials, they should be recycled or disposed of as radioactive waste. It is important to follow the instructions that come with the smoke alarm when they need to be discarded.

Space Applications: Radiation-Induced Effects

Radiation-induced spacecraft anomalies have been known since the Explorer I launch on January 31, 1958, when a Geiger counter put aboard by J.A. Van Allen suddenly stopped counting. It turned out that the counter was in fact saturated by an extremely high count rate. This led to the discovery of the Van Allen belts.

The inner belt, beginning at about 1,000 km above the surface of the Earth, contains primarily protons with energies between 10-100 MeV. The offset between Earth's geographical and magnetic axes causes an asymmetry in the radiation belt above the Atlantic Ocean off the Brazilian coast, allowing the inner belt to reach a minimum altitude of 250 km. This "South Atlantic Anomaly" is important because it occupies a region through which low-orbiting satellites spend as much as 30percent of their time. During a solar flare, which can happen anytime, the number of protons suddenly increases by more than a million.

The first spacecraft loss due to total radiation dose effects occurred unexpectedly in 1962. A satellite, Telstar, was launched just one day after a high altitude nuclear weapons test. This weapons test produced a large number of beta particles, which caused a new and very intense radiation belt that lasted until the early 1970s. Telstar and six other satellites were lost within a seven-month period after this weapons test. Telstar was well-studied and the loss was traced directly to breakdown of diodes in the command decoder due to the total radiation dose.

Radiation effects studies done at accelerators measure either total-dose effects, displacement damage, or single event effects (SEE). A SEE occurs when one ion passing through a semiconductor causes enough damage to upset the circuit in some way. In 1978, the first SEEs were observed at ground level when Intel Corporation discovered that anomalous upsets occurring in dynamic random access computer memories were being caused by alpha particles emitted from trace amounts of thorium and uranium in the materials from which the device's packages were made. It was quickly found that massive ions, protons, and neutrons could all induce SEEs.

SEEs can result in the flip of a memory bit as in the case of upsets with little other effect on the circuit, or they can be more catastrophic, causing the chip to either latchup and have to be powered down to reset the circuit, or it might burn out altogether or rupture a gate.

Biology Studies

Radiation biologists study the effect of radiation on living tissue. This field, which has direct analogies to the SEE studies, looks at the damage sustained by cells in the cosmic ray environment of space, a single event effect. These studies are necessary before we send astronauts to Mars.

Materials Studies

In 1959, it was discovered that mica, that was exposed to fission fragments, showed "nuclear tracks" when viewed under an ordinary microscope. This discovery led to a new class of nuclear physics detectors as well as a wide range of practical applications in earth science, oceanography, biology, medicine, archaeology, and space science.

By exposing thin sheets of plastic or other material to massive ions, tracks can be made through the material and then etched to make very fine "micropore" filters (30 nm to 8 μm hole size). These have been made at reactors since about 1970 and are used primarily in the biomedical field, with other applications in geophysics, meteorology, and brewing.

Another use for these sheets of plastic is to measure radon levels. By counting the number of ion tracks left in the plastic, one can measure the radon concentration.

Nuclear Medicine: Radioisotopes for Diagnosis and Treatment

One major use of radioisotopes is in nuclear medicine. Of the 30 million people who are hospitalized each year in the United States, 1/3 are treated with nuclear medicine. More than 10 million nuclear-medicine procedures are performed on patients and more than 100 million nuclear-medicine tests are performed each year in the United States alone. A comparable number of such procedures are performed in the rest of the world.

There are nearly one hundred radioisotopes whose beta and/or gamma radiation is used in diagnosis, therapy, or investigations in nuclear medicine. The most used radioisotopes were discovered before World War II using the early cyclotrons of Ernest Lawrence, with the initial applications to medicine being developed by his brother John Lawrence. Some of the most well known radioisotopes, discovered by Glenn Seaborg and his coworkers, are ^{131}I (discovered in 1938), ^{60}Co (1937), ^{99m}Tc (1938), and ^{137}Cs (1941). By 1970, 90 percent of the 8 million administrations per year of radioisotopes in the United States utilized either ^{131}I, ^{60}Co, or ^{99m}Tc. Today, ^{99m}Tc, with a half-life of 6 hours, is the workhorse of nuclear medicine. It accounts for more than 10 million diagnostic procedures a year in the United States. It is used for brain, bone, liver, spleen, kidney, lung and thyroid imaging as well as for blood-flow studies.

^{131}I, with a half-life of 8 days, is used to diagnose and treat thyroid disorders. Seaborg's mother was one of the first to benefit from the use of this radioisotope that her son had discovered. Fatally ill from hyperthyroidism, (a related condition from which her sister died), diagnosis and treatment with ^{131}I led to her complete recovery and a long life. Former President George Bush and First Lady Barbara

Bush are some notable people who were successfully treated for Graves' disease, a thyroid disease, with ^{131}I. Radioactive iodine treatment is so successful that it has virtually replaced thyroid surgery.

A very effective role for radioisotopes in nuclear medicine is the use of short-lived positron emitters such as ^{11}C, ^{13}N, ^{15}O, or ^{18}F in a process known as Positron Emission Tomography (PET). Incorporated in chemical compounds that selectively migrate to specific organs in the body, diagnosis is effected by detecting annihilation gamma rays–two gamma rays of identical energy emitted when a positron and an electron annihilate each other. These gamma rays have the very useful property that they are emitted in exactly opposite directions. When botn are detected, a computer system may be used to reconstruct where the annihilation occurred. By attaching a positron emitter to a protein or a glucose molecule, and allowing the body to metabolize it, we can study the functional aspect of an organ such as the human brain. The PET image shows where the glucose has been absorbed.

PET imaging becomes even more valuable when we can observe the functional image compared to the anatomical image. Magnetic Resonance Imaging (MRI)–originally known as Nuclear Magnetic Resonance Imaging–can provide very detailed images of the anatomy. These techniques provide researchers a better understanding of what is healthy tissue versus what is diseased.

Nuclear Medicine: Cancer Therapy

The radioisotope ^{60}Co emits gamma rays that are used to destroy cancer cells. Hundreds of thousands of Americans who suffer from cancer have been treated in this way. Every year millions of cubic meters of medical products and equipment are sterilized by irradiation worldwide. ^{137}Cs has found substantial applications as a gamma-ray source in medical therapy, similar in its use to that of ^{60}Co.

Cancer treatment with beams of massive ions directly from an accelerator has gained increasing utilization in the last decade. Unlike gamma rays, which distribute their energy equally in healthy as well as cancerous cells, massive particles such as protons or alpha particles will deposit the bulk of their energy just before they stop. If the energy is well-chosen, most of the energy will be dumped into the tumor and not into the surrounding healthy tissue. Using three dimensional water degrader columns, the shape of the tumor can be mapped out and selectively irradiated. Dedicated accelerators are now being built to continue this work at medical centres. In the United States, the Loma Linda Medical Centre in California is now operating a proton synchrotron for therapy, and a second facility is being built at Massachusetts General in Boston.

Boron Neutron Capture Therapy (BNCT) is under development for the treatment of glioblastoma multiforma, a brain cancer which afflicts some 12,500 people a year in the US alone and is almost impossible to treat by currently available means. In BNCT, boron is synthesized into compounds that are selectively taken up by cancerous cells in the brain and not by healthy ones. In subsequent irradiation by low-energy neutrons (from a reactor or accelerator), the following neutron-capture reaction occurs:

$^{10}\text{B} + \text{n}$ Æ ^{11}B Æ $^{7}\text{Li} + {}^{4}\text{He}$.

The recoiling lithium and helium nuclei have short ranges and large energy losses. These particles destroy the cancerous cells and spare the surrounding healthy tissue.

Radioactive Dating

The technique of comparing the abundance ratio of a radioactive isotope to a reference isotope to determine the age of a material is called radioactive dating. Many isotopes have been studied, probing a wide range of time scales.

The isotope ^{14}C, a radioactive form of carbon, is produced in the upper atmosphere by neutrons striking ^{14}N nuclei. The neutron is captured by the ^{14}N nucleus and knocks out a proton. Thus, we have a different element, ^{14}C. The isotope, 14C, is transported as $^{14}CO_2$, absorbed by plants, and eaten by animals. If we were to measure the ratio of ^{14}C to ^{12}C today, we would find a value of about one ^{14}C atom for each one-trillion ^{12}C atoms. This ratio is the same for all living things–the same for humans as for trees or algae.

Once living things die, they no longer can exchange carbon with the environment. The isotope ^{14}C is radioactive, and beta-decays with a half-life of 5,730 years. This means that in 5,730 years, only half of the ^{14}C will remain, and after 11,460 years, only one quarter of the ^{14}C remains. Thus, the ratio of ^{14}C to ^{12}C will change from one in one-trillion at the time of death to one in two trillion 5,730 years later and one in four-trillion 11,460 years later. Very accurate measurements of the amount of ^{14}C remaining, either by observing the beta decay of ^{14}C or by accelerator mass spectroscopy (using a particle accelerator to separate ^{12}C from ^{14}C and counting the amount of each) allows one to date the death of the once-living things.

Perhaps you have heard of Ice Man, a man living in the Alps who died and was entombed in glacial ice until recently when the ice moved and melted. The man's body was recovered and pieces of tissue were studied for their ^{14}C content by accelerator mass spectroscopy. The best estimate from this dating technique says the man lived between 3350 and 3300 BC.

The boat of a pharaoh was discovered in a sealed crypt and reassembled in a museum near the pyramids. Its wood was dated using ^{14}C to be about 4,500 years old.

Other methods of dating are used for non-living things. ^{40}K decays with a half-life of 1.3×10^9 years to ^{40}Ar which can be trapped in rocks. A potassium-argon method of dating, developed in 1966, measures the amount of ^{40}Ar arising from the ^{40}K decay and is compared to the amount of ^{40}K remaining in the rock. From the ratio, the time since the formation of the rock can be calculated.

The age of our galaxy and earth also can be estimated using radioactive dating. Using the decays of uranium and thorium, our galaxy has been found to be between 10 and 20 billion years old and the earth has been found to be 4.6 billion years old. The Universe must be older than our galaxy. Within experimental error, this estimate agrees with the 15 billion year estimate of the age of the Universe.

Neutron Activation Analysis

There are at least 50 elements occurring in nature that have radioactive isotopes with one neutron more than their stable isotopes. This means that the radioactive species can be made by neutron bombardment. This procedure is typically done in a nuclear reactor, although other neutron sources can sometimes be used. The stable nucleus absorbs one neutron and becomes a radioactive nucleus. By detecting the decay of these nuclei, which can be done with great sensitivity, one can measure the concentration of the stable element of interest in the sample.

A simple example of neutron activation analysis involves the measurement of iridium in soils. This is the measurement that led to the theory that the extinction of the dinosaurs, 65 million years ago, was caused by the impact of an asteroid or comet somewhere on Earth. An impact would produce so much debris in the air that earth would receive a large reduction in sunlight and thus fewer plants would grow. With less food available, there would be a devastating reduction in number of animals. This reduction was so severe that most species, including the dinosaurs, became extinct.

In 1979, a group of scientists reported that neutron activation analysis had shown unusual amounts of the element iridium in Italian Cretaceous-Tertiary boundary sediments. Since iridium is a metal which has very low abundance on earth, they attributed the excess iridium to a impact of a 10-kilometre diameter asteroid. This work, which melded the disciplines of physics (Nobel Laureate Luis Alvarez), geology (his son, Walter Alvarez) and nuclear chemistry (Isadore Perlman, Frank Asaro and Helen Michel), galvanized the scientific world because the concepts presented could be tested by many diverse techniques.

Since the initial report, anomalous amounts of iridium have been found at the Cretaceous-Tertiary boundary in over 100 sites worldwide. Many experiments havè confirmed its impact origin.

Industrial Applications

The applications of radioisotopes in industry are numerous. Many types of thickness gauges exploit the fact that gamma rays are attenuated when they pass through material. By measuring the number of gamma rays, the thickness can be determined. This process is used in common industrial applications such as:

1. The automobile industry–to test steel quality in the manufacture of cars and to obtain the proper thickness of tin and aluminum
2. The aircraft industry–to check for flaws in jet engines
3. Construction–to gauge the density of road surfaces and subsurfaces
4. Pipeline companies–to test the strength of welds
5. Oil, gas, and mining companies–to map the contours of test wells and mine bores, and
6. Cable manufacturers–to check ski lift cables for cracks.

The isotope ^{241}Am is used in many smoke detectors for homes and businesses (as mentioned previously), in thickness gauges designed to measure and control metal foil thickness during manufacturing processes, to measure levels of toxic lead in dried paint samples, and to help determine where oil wells should be drilled.

The isotope ^{252}Cf (a neutron emitter) is used for neutron activation analysis, to inspect airline luggage for hidden explosives, to gauge the moisture content of soil and other materials, in bore hole logging in geology, and in human cervix-cancer therapy.

In addition, there are manifold uses in agriculture. In plant research, radiation is used to develop new plant types to speed up the process of developing superior agricultural products. Insect control is another important application; pest populations are drastically reduced and, in some cases, eliminated by exposing male insects to sterilizing doses of radiation. Fertilizer consumption has been reduced through research with radioactive tracers. Radiation pellets are used in grain elevators to kill insects and rodents. Irradiation prolongs the shelf-life of foods by destroying bacteria, viruses, and molds.

The useful application of radioisotopes extends to the arts and humanities. Neutron activation analysis is extremely useful in identifying the chemical elements present in coins, pottery, and other artifacts from the past. A tiny unnoticeable fleck of paint from an art treasure or a microscopic grain of pottery suffices to reveal its chemical makeup. Thus the works of famous painters can be "fingerprinted" so as to detect the work of forgers.

Neutron scattering has proved to be a valuable tool for studying the molecular structure and motion of molecules of interest to manufacturing and life processes. Accelerators and reactors produce low-speed neutrons with wavelength appropriate to "see" structures of the size of magnetic microstructures and DNA molecules. Neutrons can penetrate deeply into bulk materials and use their magnetic moment or strong interaction forces to preferentially scatter from magnetic domains or hydrogen atoms in long chain nucleosomes. Neutrons are also used in materials surface and interface studies taking advantage of their reflectivity properties. Intense sources of neutrons include: the IPNS at Argonne National Laboratory in Illinois and LANSCE at Los Alamos National Laboratory in New Mexico.

Removal of Land Mines

Wars have been fought around the globe, and land mines were used extensively by the battling parties. There may be up to 100 million abandoned mines just waiting to maim and kill unsuspecting civilians. Techniques similar to airport weapons detectors have been considered for finding and neutralizing these mines without loss of life.

A promising technology involves nuclear quadrupole resonance. Nitrogen, a common component in explosives, has a nucleus with an ellipsoidal shape. Depending on what kind of crystalline structure the nitrogen nuclei find themselves in, their non-sphericity produces a unique set of very narrowly spaced energy levels that is characteristic of the crystalline solid itself. An explosive compound can therefore be identified by the subtle effect of its constituent nitrogen atoms. This technique could also differentiate explosives from scrap metals in the ground. More details may be found in the reference at the end of this chapter.

Radioisotope Power Generation

Long lived power sources are needed for equipment that is too remote or inaccessible for replacement. By choosing a radioactive element with a long half life, we can create a long lived power source. The appropriate element should:

1. Produce weakly penetrating radiation that can be easily shielded
2. A specific power of at least 0.2 kW/kg
3. Have good corrosion resistance
4. Be insoluble in water
5. Be made of reasonably available material.

Among the transuranium elements, oxides of the alpha-emitting nuclides ^{238}Pu (t1/ 2 = 87.7 years) and ^{244}Cm ($t_{1/2}$ = 18.1 years) are useful fuels. A few grams to kilograms of such nuclides, in appropriately shielded containers, provide intense sources of heat with power levels up to hundreds of watts, since the alpha particles are stopped very easily and their decay energy converted into thermal energy. Using thermoelectric devices without moving parts, it is possible to convert the resultant heat flow into usable electricity.

Such power sources are small, lightweight, and rugged. One of their uses is in the SNAP (Space Nuclear Auxiliary Power) units used to power satellites, or more importantly, to power remote sensing instrument packages. SNAP sources (fueled by ^{238}Pu) served as the power sources for instrument packages on the five Apollo missions, the Viking unmanned Mars lander, and the Pioneer, Voyager and Cassini probes to Jupiter, Saturn, Uranus, Neptune, Pluto, and beyond.

In addition to space applications, radionuclide power sources have been used as terrestrial sources of energy wherever compact and long-lived sources, not requiring maintenance, are needed. In the early 1970s, ^{238}Pu batteries were used as the power sources for cardiac pacemakers. Over 3,500 units were implanted and most remain functioning. Due to a lower cost and easier construction, lithium batteries have largely supplanted the plutonium.

Nuclear Weapons

Without a doubt, the development of nuclear weapons is one application of nuclear science that has had a significant global influence. Following the observation of fission products of uranium by Hahn and Strassmann in 1938, a uranium fission weapon became possible in the eyes of a number of nuclear scientists. It was Albert Einstein who signed a letter to Franklin Roosevelt, President of the United States, and alerted him to the potential development of a nuclear weapon. Thus began the Manhattan Project, resulting in production of the first nuclear weapons.

In 1945, a bomb using the fission of ^{235}U was dropped on Hiroshima, while a bomb using the fission of ^{239}Pu was dropped on Nagasaki. The awesome destruction that was produced by two nuclear devices changed the face of warfare forever. Not only did these two explosions end the most destructive war in history, but the possessor of these weapons of mass destruction seemed invincible to any adversary.

The legacy of nuclear weapons remains in the soil around us, where we can still measure minute amounts of fallout from the atmospheric nuclear tests of the 1950s and 1960s. The weapons need constant maintenance because radiation damage affects the materials and triggering devices that must act together to make the weapon work. Furthermore, the isotope tritium, which is used in weapons, naturally decays. Storage and disposal of the weapons is technically difficult but probably a manageable problem. However, it is a daunting political and social problem. It is a challenge to maintain these weapons, to significantly reduce their numbers, to clean up the waste, and to insure that these weapons never will be used again.

Nuclear weapons are with us to this day and could be with us in the future. Even if all conflicts among nations end, some nuclear weapons might be retained. Earth is vulnerable to impacts from comets and asteroids. Some scientists have proposed that nuclear weapons could possibly be used to deflect them from hitting our planet.

Agricultural Applications

Radioisotopes can be used to help understand chemical and biological processes in plants. This is true for two reasons: 1)radioisotopes are chemically identical with other isotopes of the same element and will be substituted in chemical reactions and 2)radioactive forms of the element can be easily detected with a Geiger counter or other such device.

Example:

A solution of phosphate, containing radioactive phosphorus-32, is injected into the root system of a plant. Since phosphorus-32 behaves indentically to that of phosphorus-31, the more common and non-radioactive form of the element, it is used by the plant in the same way. A Geiger counter is then used to detect the movement of the radioactive phosphorus-32 throughout the plant. This information helps scientists understand the detailed mechanism of how plants utilized phosphorus to grow and reproduce.

Food Irradiation

Food irradiation is a method of treating food in order to make it safer to eat and have a longer shelf life. This process is not very different from other treatments such as pesticide application, canning, freezing and drying. The end result is that the growth of disease-causing microorganismns or those that cause spoilage are slowed or are eliminated altogether. This makes food safer and also keeps it fresh longer.

Food irradiated by exposing it to the gamma rays of a radioisotope — one that is widely used is cobalt-60. The energy from the gamma ray passing through the food is enough to destroy many disease-causing bacteria as well as those that cause food to spoil, but is not strong enough to change the quality, flavor or texture of the food. It is important to keep in mind that the food never comes in contact with the radioisotope and is never at risk of becoming radioactive.

The FDA has approved the irradiation of several food catagories, but irradiation is most widely used on spices, herbs and dehydrated vegetables. Since these food items are grown in or on the ground it is clear to see that they are at risk for being exposed to naturally occuring pests such as bacterias, molds, fungi, insects, and rodents. It is impossible to harvest and package these items without having some contamination from these naturally occuring pathogens. Irradiation of this material can help to ensure that the final product you purchase is pest free.

Some meats are irradiated. Pork, for example, is irradiated to control the trichina parasite that resides in the muscle tissue of some pigs. Poultry is irradiated to eliminate the chance of foodborne illness due to bacterial contamination.

Irradiation of certain foods also have additional benefits. Since the energy passing through the food can disrupt cellular processes (this is the mechanism for destroying microorganisms) it also can halt the cellular processes that lead to the sprouting or ripening of foods. Potatoes and onions are irradiated to retard their sprouting. Fruits and vegetables are irradiated to slow down the ripening process. In this way, delicate fruits won't reach their peak ripeness before they arrive at the supermarket.

Food irradiation sounds like a wonderful use of nuclear chemistry principles. Although this process is routinely used in Europe, Canada, and Mexico, the United States has been a little more hesitant to adopt food irradiation. This is due mainly to the public's perceived fear and limited understanding of nuclear science. For example, although the FDA has approved the irradiation of poultry, the industry hesistates to adopt the process because they are afraid of a negative response from consumers. With recent public education, however; many people are learning to appreciate and value it's usefullness.

INDUSTRIAL APPLICATIONS

Ion Implantation in Semiconductor

Ion implantation is a process in which ions of a material can be implanted to another solid, thereby changing the physical properties of the solid. Semiconductors like silicon can alter the conductivity by introducing a small quantity of dopant atoms like boran, arsenic phosphorus. It is important to accurately control the number of implanted dopants and to place them at a desired depth. The ion implantation equipment typically consists of an ion source, an accelerator and a target chamber. The ions of the desired element are produced and these are accerelerated in the electric field to a high energy. A magnet selects the desired beam and directs it towards the target, which is the material to be implanted.

Material Analysis

Isotope Dating

Radioactive isotope dating is a sophisticated method of determining the age of a material. This is done by making a comparison between the abundance ratio of a radioactive isotope to a reference isotope. The radioactive isotope C-14 has a half life period of 5,730 years. The half-life of a radioactive isotope is the time it takes for half the isotopes in the sample to decay away. So, for a sample of C-14, after 5,730 years only half of the original number of 14C will remain, and after 11,460 years, only one quarter of the 14C remains. Very accurate measurements of the amount of carbon-14 in a material can be done either by observing the beta decay of 14C or by accelerator mass spectroscopy (using a particle accelerator).

Accelerator Mass Spectroscopy

The basic principle in accelerator mass spectrometry is to separate various atoms according to their charge-to-mass ratio. An ion source bombards the sample to be dated with a cesium beam that negatively ionizes and releases the carbon atoms from the sample. Impurities are also present. A magnet and slit arrangement gets rid of some of the impurities before the negative beam is injected into an accelerator with a field of millions of volts. The accelerated ions hit a thin carbon foil that strips off electrons, and in so doing, destroys all molecular species. The carbon atoms, now positively charged ions, continue to be accelerated. On exiting the accelerator, the ions are passed through a mass spectrometer and various filters that separate the particles according to their mass. Finally, the ions slow down to a stop in a gas ionization detector according to their charge. The identity of individual ions can be determined from how the ions slow down. For example, carbon-14 slows down more slowly than nitrogen-14, so those ions of the same mass can be distinguished from one another. Once the charges are determined, the detector can tell to which element each ion belongs and counts the desired isotope as a ratio of a more abundant isotope-carbon-14.

Sterilization

Particle accelerators are use for medical sterilization, food preservation, purifying drinking water and removing noxious substances. There is lot of food born illness erupting which could be mitigated through the sterilization of food using accelerators, but this method is still not widely accepted by the general public. Irradiation is the most effective method of eliminating the deadly bacteria in raw meat. It can also significantly reduce levels of other pathogens, including Listeria, Salmonella, Shigella, Campylobacter and others. While there is no single technique to cure all food problems, irradiation is accepted as both safe and effective. Linear accelerators can be used for irradiation. There are also some sterilization processes which accelerators can perform. An accelerator is an alternative to the autoclave to sterilize surgical instruments and laundry. Also, the preservation of food stuff for animals and production of fertilizer for crops is commonly done, as these are perceived to be sufficiently remote from human ingestion, so that the public is not afraid of a radiation hazard.

Microstructure Lithography

X-rays from light sources can be focused to such a fine point that they can cut out tiny defects such as machine components from silicon or plastic with incredible precision on a scale of thousandths of a millimeter. They can also be used to etch patterns in microchips

Explosives

Particle accelerators produce charged particles and accelerate them into a high energy beam; particle beams may be used for many different applications. In the 1950s scientists who were doing research in the material sciences found that particle accelerators can be used in producing heavy metals such as Uranium-233 and Plutonium-239. While studying cosmic rays it was discovered that high energy particles can be used to produce neutrons. This enabled the use particle accelerators for the production of fission material used in explosives. Fission material such as uranium-233 is produced from fertile material thorium-232; when an atom of thorium-232 absorbs a neutron, Uranium-233 is produced. Similarly plutonium-239 can be produced from Uranium-238 by absorbing a neutron.

Transmutation of Nuclear Waste

Apart from developing fission material for explosives, particle accelerators can also be used for treating nuclear waste. Nuclear waste from nuclear reactors is stored in storage pools, but nowadays the storage pools are filled completely and the radioactive materials in the waste have half-life period of million years. Since there is no way to dispose of the nuclear waste without causing a threat, a new concept called ADTT (Accelerator Driven Transmutation Technology) has been developed. It can reduce the half-life period of the radioactive waste to a few centuries which would be helpful in storing the waste. The conceptual basis is to use neutrons to transmute the particles of radioactive elements such as Plutonium and Neptunium to products with a lower half-life. Accelerator based conversion can also be used to destroy plutonium from weapons and reactor fuel sources by producing products with lower half-life. This process can burn up to 98 per cent of plutonium-239 and 90 per cent of plutonium isotopes from explosive material.

Index

H

K

L

MC

O

P

R

S

T

V